Supplement
to
Human Nutrition

Jack Bateman

Miami Dade Community College

KENDALL/HUNT PUBLISHING COMPANY
4050 Westmark Drive   Dubuque, Iowa 52002

Image of sunflower seeds courtesy of Corel

Grain, peas, nuts and wheat images © 2001 PhotoDisc, Inc.

Contents

CHAPTER 1   1
INTRODUCTION TO NUTRITION

Chapter 2
Digestion

31

Chapter 3
Carbohydrates

49

CHAPTER 4
LIPIDS

73

CHAPTER 5
PROTEINS

91

Chapter 9
Fat Soluble Vitamins

157

Chapter 10
Minerals

165

Preface

The nutrition course that we tech at Miami-Dade Community College is a course in the science of nutrition. This implies that this discipline will involve nutrition at the molecular, cellular level. For this reason it is necessary that the student have a working understanding of the composition of food molecules as well as their molecular metabolic fates. As a professor of nutrition for over 25 years, I have always found it necessary to supplement the text book with many of my own cartoons, tables, graphs and drawings in order to adequately illustrate these molecular concepts. The constant urging of many of my devoted students have inspired me to put this supplement together. I hope it is helpful in guiding you through this course.

Acknowledgments

I am very grateful to my devoted student, Maria Ricker, for providing me with her thorough, typed class notes, many of which appear in this book. I would like to thank Dr. Kenneth Pelletier, PhD., University of California, College of Medicine, San Francisco, CA., for teaching me the fundamentals of molecular nutrition. I would also like to thank Dr. Leo Galland, M.D. for the many inspirational lectures on fatty acid metabolism and holistic medicine, and Dr. Don Clark, Phd., Miami-Dade Community College, for his constant friendship and academic stimulus. A special thanks to my wife, Dinorah, for her constant infusion of medical research and her astute contributions to the nutrition projects.

CHAPTER 1
INTRODUCTION TO NUTRITION

Class Notes

Basic Nutrient Classes

- Protein
- Carbohydrate
- Lipids
- Vitamins
- Minerals
- Water

The Energy Nutrients

- Cell membranes
- Nucleus
- Rough endoplasmic reticulum
- Smooth endoplasmic reticulum
- Golgi Complex
- Mitochondria
- Lysosomes

Molecular Genetics and Nutrition

- DNA—genes
- DNA → RNA → Protein
- Gene Control
- Mutations and disease

States of Nutritional Health

- Healthy vs. Optimal
- Optimal nutritional status
  - Optimal nutrient consumption

- Optimal metabolic function
- Optimal immune function
- Optimal health
- Compromised nutritional status
  - Overnutrition
  - Biochemical lesions—sub-clinical symptoms
  - Clinical lesions—clinical symptoms

Basic Food Groups

Food Pyramid

- USDA
- Alternative pyramids

Nutrient Density vs. Exchange System

Typical American Diet vs. U.S. Government Recommended Diet

RDAs and RDIs

Safety of the Food Supply

Food Labeling

Nutritional Research

- Scientific Journals—peer reviewed articles vs. popular magazines
- Computer searches on Pubmed/Medline
- Research types
- Research design
- Examples of research

Fate of Food

- Ingestion
- Digestion
- Absorption
- Circulation
- Metabolism
  - Catobolism—ATP
    - ATP cycle
  - Anabolism—the compound of life

The Atoms of Nutrition

Simple Structure of Nutrients

Basic Chemical Reactions of Nutrition

- Oxidation—Reduction
- Dehydration of condensation synthesis
- Hydrolysis or digestive degradation
- Phosphorylation
- Acid-Base reactions

Top Ten Causes of Death—Relationship to Nutrition

Lipids

- Triglycerides
  - Glycerol and fatty acids
  - Stored in adipose tissue—secondary energy source in most cells
  - Synthesized in the *Fatty Acid Cycle*
- Cholesterol
  - Made in a spin-off of the fatty acid cycle
  - Cannot be catabolized
  - Used in making cell membranes
- Phospholipids
  - Used in making cell membranes

Nutrients Classes

Protein

Carbohydrates

Lipids

Vitamins

Minerals

Water

Macronutrients

Energy

Protein

Carbohydrates

Lipids

Fiber

Micronutrients

Water Soluble Vitamins

Fat Soluble Vitamins

Major Minerals

Trace Minerals

Energy Sources—The Kcal Nutrients

Carbohydrates—4 Kcals/g

Proteins—4 Kcals/g

Lipids—9 Kcals/b

Alcohol—7 Kcals/g

Macronutrients

Energy

Proteins
Carbs } Kcal Nutrients—Only 3 from which we get calories
Lipids

Fiber—no kcals, passes through digestive tract to assist bowel movement.
The nutrient classes vitamins, minerals, and water have no kcals.

Proteins

These molecules are polymers (many parts) of amino acids

Proteins are composed of 50–500 of amino acid residues

Amino Acids

- 20 Different Amino Acids in Human Protein
  - 9 Essential—body does not produce; body needs or it will die
  - 11 Non Essential—body makes them.

The Worker Molecules

- Hormones—Chemical Messengers
- Enzymes—Proteins that facilitate chemical reactions without being hanged in the process.
- Antibodies—Protein molecules that neutralize foreign molecules (antigens).
- Receptors—Specific proteins—shaped to receive only specific molecular messages.
- Transporters— ex. Hemoglobin—transports oxygen in the blood stream.

Carbos—No. 1 Energy Nutrient

- Sugars—Glucose—must have in the blood
- Starch
- Glycogen
  - Made and stored in liver and muscles
  - Our energy reserve; provides energy for flight or fight response

Lipids

- Fats and Oils

- Triglycerides
- Cholesterol
- Phospholipids—Cell Membranes (Phos + Chol Make up Cell Membranes)
- No. 1 Lipid in Body

Energy Sources

- Carbo—4Kcal/g
- Protein—4Kcal/g
- Fats—9Kcal/g
- Alcohol—7Kcal/g (Not a Nt / Robs Calories from C, P & Fs)

Water Soluble Vitamins (molecular structure attracts water molecules)

- C—Ascorbic acid
- B-1—Thiamin
- B-2—Riboflavian
- B-3—Niacin
- B-5—Pantothenic acid
- B-6—Pyridoxal
- B-9—Folacin
- B-12—Cyanocobalamin and Biotin

Fat Soluble Vitamins

- A—Retinol
- D—Cholecalciferol
- E—D-Alphatocopherol
- K—Phylloquinone

Proteins

Polymers of Amino Acids

- 50–500 amino acid residues
  7500–75,000 Da

The Worker Molecules

Enzymes

Antibodies

Hormones

Receptors

Transporters

Signaling

Amino Acids

9 Essential

11 Non-Essential

Classification

    Acidic Amino Acids

    Basic Amino Acids

    Neutral (Hydrophobic) Amino Acids

    Sulfur Amino Acids

    Polar Amino Acids

Carbohydrates

No. 1 Energy Nutrient

Sugars—Glucose

Starch

Glycogen

Lipids

Fat and Oils—Triglycerides

Cholesterol

Phospholipids—Cell Membranes

Cellular Structure

Almost everything inside a cell is also made up of cell membrane materials. Organelles (little organs), little parts inside the cell, are mostly made of cell membrane material.

External Cell Membrane = Plasma Membrane (the outside of the cell, made up of phospholipids)

Nucleus is made up of a double membrane layer, has little holes called nuclear pores.

Nuclear Pores are proteins that can open or close (proteins are functional, allow material in and out of nucleus).

- The nucleus contains the DNA. DNA makes RNA, which directs making of Protein.
- RNA is made from the DNA template in Nucleus. RNA travels out of nucleus through Nuclear Pores to Rough Endoplasmic Reticulum.

Ribosomes are Protein factories. Their job is to manufacture Protein.

- Ribosomes assemble across a piece of messenger RNA, then read the RNA code to translate it into Protein. As it slides across messenger RNA it reads the base sequence of the RNA and turns this message into Protein. This process occurs in the Rough Endoplasmic Reticulum, which is attached to Nucleus and External Cell Membrane.

How does a Gene or DNA (gene = DNA) express itself?

- DNA directs the making of Protein (via RNA)
- The Central Dogma— DNA → RNA → Protein

Smooth Endoplasmic Reticulum—contains enzymes which manufacture Complex Carbs (glycogen) and Complex Lipids (phospholipids)

Mitochondrion—Powerhouse of the cell

- Organelles that produce ATP
- ATP factory in cell where Catabolism takes place

F.Y.I.—When you do aerobic exercise on a regular basis, you produce more Mitochondria. The more Mitochondria you produce, the more ATP you produce. Aerobic exercise stimulates the production of growth factors that encourage the production of protein in muscles as well as mitochondria.

Golgi Body—bagging factory, produces Vesicles that store cell products.

Vesicles (a bag made of phospholipids)—like little Ziploc bags that store cell products; ball made up of Cell Membrane.

Lysosome (Lys=break apart/some=body)—breaks apart foreign materials.

- Vesicles that contain specific digestive enzymes
- Cellular organelles; membrane enclosed sacks of degradative enzymes

Nutrient Density is a powerful tool in assessing nutrient quality. It is an effective method of comparing the quality of one food to another. This fills the quality void in the exchange system. Throughout this supplement, foods will be compared with respect to a specific nutrient by nutrient density (a nutrient divided by the energy in kcals). In calculating nutrient density of a given food, the most effective method is to convert the vitamins and minerals into percents of the RDAs or RDIs, then sum these and divide by the energy in kcals. This is a labor intensive calculation, but rich in information.

Nutrient Density

Definition:

$$\text{Nutrient Density (ND)} = \frac{\text{Essential Nutrients}}{\text{Energy (Kcals)}}$$

Practical Definition:

$$\text{ND} = \frac{\text{Sum of Vitamins and Minerals*}}{\text{Energy (Kcals)}}$$

*Best method is to calculate the sum of vitamins and minerals as percents of the RDAs.

Application:

$$\text{ND} = \frac{\text{Single Nutrient}}{\text{Energy (KCals)}}$$

Example: **3.4 oz of Grouper:**

$$\text{ND}_{\text{protein}} = \frac{\text{25 grams of Protein}}{\text{101 Kcals}} = .25$$

The Fate of Food

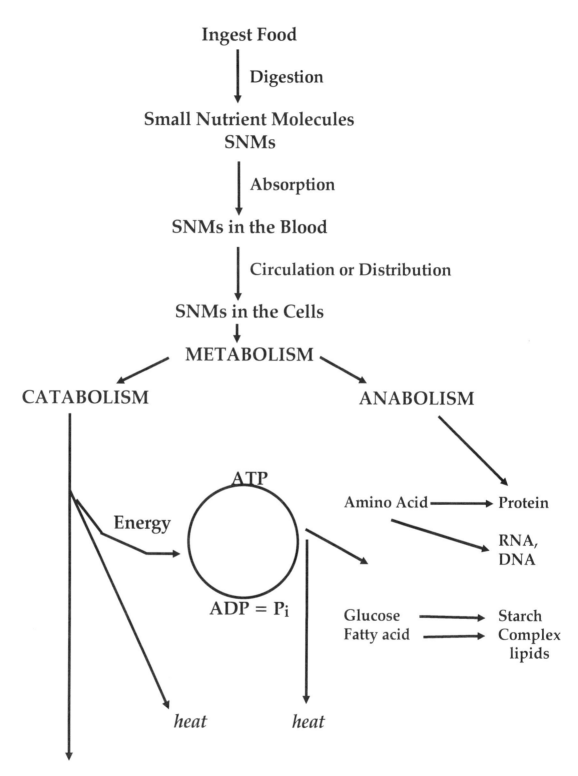

Ingest Food

Digestion

Small Nutrient Molecules
SNMs

Absorption

SNMs in the Blood

Circulation or Distribution

SNMs in the Cells

METABOLISM

CATABOLISM

ANABOLISM

ATP

Energy

ADP = P_i

Amino Acid $\longrightarrow$ Protein

RNA,
DNA

Glucose $\longrightarrow$ Starch
Fatty acid $\longrightarrow$ Complex
lipids

heat   *heat*

Carbon Dioxide and Water

The ion magnesium (MG++) is a coenzyme and absolutely necessary for the synthesis and function of ATP.

Ingest—the process of eating food. Chewing (mastication) breaks food into smaller pieces so enzymes can get to it and begin process of *Digestion*. Digestion breaks up food into *Small Nutrient Molecules* (SNMs). These small nutrient molecules are in our gut (digestive tract) and are not ours yet, i.e. we need to absorb them before we own them. *Absorption*— we have to absorb SNMs to make them ours. During absorption the SNMs pass across the membranes into the blood stream. *Snms in Blood Stream*—Now we have SNMs in the blood, which is the way we *Circulate and Distribute* the SNMs throughout the body. The goal is to get the SNMs out to the Cells of the body through Circulation and Distribution in the body. Now we have SNMs in Cells. In Cells is where we do Metabolism. *Metabolism* is the term we give to the chemical reactions that go on inside the Cell. We break Metabolism into two categories: Catobolism and Anabolism. *Catobolism is* the process of making ATP. The No. 1 SNM we use in Catabolism is Glucose. Glucose is our energy nutrient. We start with Glucose and break it down into *Carbon Dioxide and Water*. The energy that is extracted from the small nutrient molecules is used to make ATP (the high energy molecule of life). *Anabolism* is the process of making the compounds of life. Anabolism — Amino Acid → protein, RNA, DNA. Glucose   → Compex Carbs. Fatty Acids → Complex Lipids.

Nutritional Research Types

Epidemiological Studies—Uncontrolled

Case-Control Studies—Participants Closely Matched

Animal Studies—Good for Modeling a Disease

Human Intervention (Clinical) Trials—
Uses Strict Inclusion/Exclusion Criteria

Nutritional Research Design

Sample Size Chosen to Meet Statistical Need for Significance

Control Group Used

Placebos Used

Single Blind vs. Double Blind

Ann Intern Med 2000 Aug. 15;133(4);245-52   Related Articles, Books, LinkOut
Annals of Internal Medicine

Duration of symptoms and plasma cytokine levels in patients with the common cold treated with zinc acetate. A randomized, double-blind, placebo-controlled trial.

Prasad AS, Fitzgerald JT, Bao B, Beck FW, Chandrasekar PH

Department of Medicine, Wayne State University, University Health Center, Detroit, Michigan 48201, USA.

BACKGROUND: Adults and children in the United States get two to six colds per year. Evidence that zinc is effective therapy for colds is inconsistent. OBJECTIVE: To test the efficacy of zinc acetate lozenges in reducing the duration of symptoms of the common cold. DESIGN: Randomized, double-blind, placebo-controlled trial. SETTING: Detroit Medical Center, Detroit, Michigan. PATIENTS: 50 ambulatory volunteers recruited within 24 hours of developing symptoms of the common cold. INTERVENTION: Participants took one lozenge containing 12.8 mg of zinc acetate or placebo every 2 to 3 hours while awake as long as they had cold symptoms. MEASUREMENTS: Subjective symptom scores for sore throat, nasal discharge, nasal congestion, sneezing, cough, scratchy throat, hoarseness, muscle ache, fever, and headache were recorded daily for 12 days. Plasma zinc and proinflammatory cytokine levels were measured on day 1 and after participants were well. RESULTS: Forty-eight participants completed the study (25 in the zinc group and 23 in the placebo group). Compared with the placebo group, the zinc group had shorter mean overall duration of cold symptoms (4.5 vs. 8.1 days), cough (3.1 [95% CI, 2.1 to 4.1] vs. 6.3 [CI, 4.9 to 7.7] days, and nasal discharge (4.1 [CI, 3.3 to 4.9] vs. 5.8 [CI 4.3 to 7.3] days), and decreased total severity scores for all symptoms (P<0.002, test for treatment x time interaction). Mean changes in soluble interleukin-1 receptor antagonist level differed nonsignificantly between the zinc group and the placebo group (difference between changes, -89.4 pg/mL [CI, -243.6 to -64.8 pg/mL]). CONCLUSION: Administration of zinc lozenges was associated with reduced duration and severity of cold symptoms, especially cough. Improvement in clinical symptoms with zinc treatment may be related to a decrease in proinflammatory cytokine levels; however, in this study, the observed differences between changes in cytokine levels in zinc and placebo recipients were not significant.

Zinc Intervention Trial vs. The Onset of a Cold

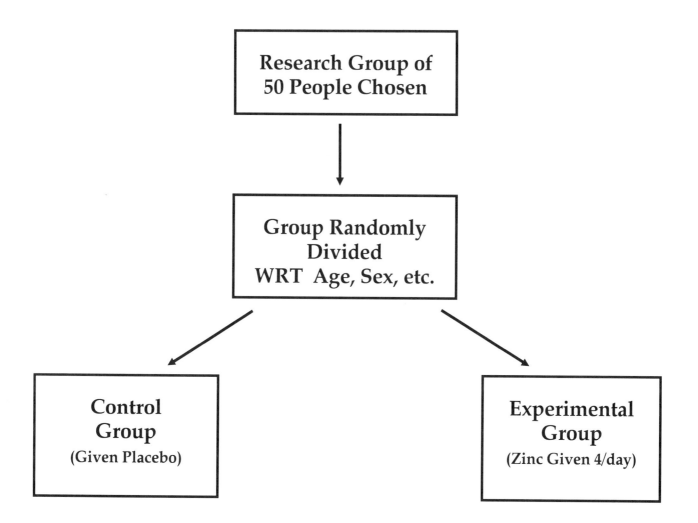

Results

- Reduced duration and symptoms of colds

Chelation

- Wrap organic molecule around a metal ion (minerals)
- Keeps it from bonding with undesirable molecules
- Vitamin C is a chelating agent for calcium and other minerals

---

Atoms found in nutrients for the following bonds—easy way to remember
HONC (like honk) 1,2,3,4

H –	1 bond	H-H	Hydrogen
-O-	2 bonds	H-O H	Water
-N-	3 bonds	H-N-H H	Ammonia
-C-	bonds	H H-C-H H	Methane

This hydrocarbon is fairly non-reactive. In order to make it reactive, we need to add an Oxygen or a Nitrogen. When we introduce O or N we make hydrocarbons reactive.

Ethanol—Primary Alcohol

$$
\begin{array}{ccc}
\text{H} \ \text{H} & & \text{H} \\
| \ \ \ | & & | \\
\text{H-C-C-O-H} & \longrightarrow & \text{H-C-C=O} \\
| \ \ \ | & & | \\
\text{H} \ \text{H} & \searrow & \text{H-H} \\
& 2\text{H} &
\end{array}
$$

Two of these H's come off of the carbon atom that has the oxygen attached to form an aldehyde.

$$
\begin{array}{ccc}
\text{H} & & \text{H} \\
| & & | \\
\text{H-C-C=O} & \longrightarrow & \text{H-C-C=O} \\
| \ \ | & +\text{O} & | \ \ \ \ \backslash \\
\text{H} \ \text{H} & & \text{H} \ \ \ \text{O-H}
\end{array}
$$

Oxidation is defined by the addition of alcohol-2H or +O

Alcohol → Aldehydes → Acids

Reduction +2H or -O

$$
\begin{array}{ccc}
\text{H} \ \ \text{H} \ \ \text{H} & & \text{H} \ \ \ \ \text{H} \\
| \ \ \ | \ \ \ | & & | \ \ \ \ \ \ | \\
\text{H-C—C—C-H} & \longrightarrow & \text{H-C—C—C-H} \\
| \ \ \ | \ \ \ | & \searrow & | \ \ \ \| \ \ \ | \\
\text{H} \ \text{OH} \ \text{H} & 2\text{H} & \text{H} \ \ \text{O} \ \ \text{H}
\end{array}
$$

This is a secondary alcohol   This is a ketone

Aldehyde, ketone, primary alcohol, and secondary alcohol groups are all found in carbohydrates.

$$
\begin{array}{c}
\text{H} \\
| \\
\text{H—C—OH} \\
| \\
\text{H—C—OH} \\
| \\
\text{H—C—OH} \\
| \\
\text{H}
\end{array}
$$

Glycerol

Glycerol is the backbone of the fat molecule, triglyceride

All sugars are either aldehydes or ketones. They are important because we turn glucose into these molecules.

$$
\begin{array}{c}
\text{H} \\
| \\
\text{H—C—OH} \\
| \\
\text{C=O} \\
| \\
\text{H—C—OH} \\
| \\
\text{H}
\end{array}
\qquad
\begin{array}{c}
\text{H—C=O} \\
| \\
\text{H—C—OH} \\
| \\
\text{H—C—OH} \\
| \\
\text{H}
\end{array}
$$

Simplest of all sugars

DHA or Dihydroxy Acetone   GAL or Glyceraldehyde

Rules of Metabolism

- Primary Alcohol → Aldehyde → Acid
  - Primary alcohol oxidizes to aldehyde which oxidizes to acid
- Secondary Alcohol → Ketone
  - Alcohol oxidizes to ketone

Must Recognize "G" Sugars

Top 10 Causes of Death in the U.S.—1997

Heart Disease	31.4%
Cancer	23.3%
Stroke	6.9%
Lung Disease	4.7%
Accidents	4.1%
Pneumonia/Flu	3.7%
Diabetes	2.7%
Suicide	1.3%
Kidney Disease	1.1%
Liver Disease	1.1%

Basic Four Food Group Plan

Dairy Group

- 2 servings/1000 kcals*

Meat and Meat Substitute Group

- 2 servings/1000 kcals

Fruit and Vegetable Group

- 4 servings/1000 kcals

Grains Group

- 4 servings/1000 kcals

*1 serving is 2 to 3 ounces

Nutritional Analysis—Phase I

Introduction and General Instructions

In phase I of the Nutritional Analysis you will collect nutritional data on the foods that you eat, total each nutrient parameter each day, and average the data. The computer program for nutritional analysis will automatically do these arithmetic calculations for you. You will compare these data to established norms. This will enable you to suggest ways in which to change your nutritional regimen, so that you can accomplish an improved nutritional status.

The first step is to record everything you eat and drink for seven days, recording both **what you are consuming and how much.** It is vital that you make every effort to accurately record the proper amounts of each item. After four days of recording foods, you should visit the Natural Science Computing Center (NSCC) in room 3319, where you will enter your list of foods and drinks for each day into the computer for analysis. You will print out

> (1) the list of foods,
> (2) the summary of nutrients for each day,
> (3) the seven day average summary, and
> (4) the food pyramid.

(Note: there are many other items in the computer menu which you could print out. You are not required to print out any additional pages. *If you are given a personal profile printout, throw it away!*)

Calculation of a Set of Ideal Values

Ideal Weight

A set of ideal values will be calculated, based on your personal lifestyle and physiological needs. You must first establish whether you are at or about your ideal weight. This can be accomplished by using the "Body Mass Index" chart on the back cover of the text, *Understanding Nutrition.* Find your height (without shoes) on the height scale and the BMI value you choose between the values of 19 and 24. Use your blouse or shirt size as a guide, i.e. small = 19–20, medium = 21–22, large = 23–24. **Where the height row crosses the BMI column is your ideal weight. You must show the comparison between your ideal weight and your real weight by listing the numbers on a page titled** *Ideal Weight.*

Calculation of Ideal Energy (Kcals)

Since the amount of food kcals you consume is controlled by the hypothalamus in the brain, you **CAN ASSUME** that **your average kcals consumption is ideal for you, if you are not overweight!** If you are overweight, then you need to decrease your average kcal by the percent you are overweight to obtain an ideal energy value. *Do not use the macronutrient values on the personal profile printout!*

Ideal Macronutrient Values

Using your ideal energy value, calculate an ideal kcal distribution according to the following percentage ranges and display your data in the following table:

Nutritional Parameter	Ideal Values	Mean Values	Differences
Energy	your average value		
Protein	calc. value*		
Carbohydrate	calc. value		
Fat	calc. value		
Fiber	17g/1000kcal 25g minimum		

A typical calculation is shown below:

$$\text{* ideal kcal} \times .12 \times \frac{1 \text{ g protein}^1}{4 \text{ kcal}} = \text{ideal protein}$$

[1] Use the conversion lg of carbohydrate/4 kcal and lg of fat/9 kcal.

The Comparison Table

The ideal values for *Energy, Protein, Carbohydrates, Total Fat*, and *Fiber* should be displayed in a table showing a comparison with your average values and the difference between the two values as shown above.

The Food Pyramid

You will get a computer printout that shows the comparison between your distribution and the recommended on a drawing of the food pyramid. Your servings of nutrient classes will be shown on one side of the pyramid with the recommended servings down the middle.

Micronutrients

Average vitamin and mineral values that are displayed in the seven-day average printout (be sure to list all of them) will be compared to the **RDAs** (last page of the textbook) and tabled appropriately showing the differences. There are no methods that have been developed for calculating ideal values for vitamins and minerals. We are still learning about the complex roles that each of these plays in the web of metabolic pathways.

Micronutrient	RDA	Mean Values	Differences

Discussion (labeled sections are in bold)

You will write a discussion in which you explain how you chose an **ideal weight and an ideal energy intake** (see directions under calculation of ideal energy). Discuss the **main sources of your kcals from your food lists**, and what foods actually contributed most to you kcal intake. **Discuss your serving distribution relative to the recommended serving ranges displayed on** your **food pyramid**. Are your main sources of kcals from quality foods? Does your pyramid distribution match the recommended ranges? What are the possible consequences? Identify the parameters in your **macronutrient consumption** that are in **excess** of your ideal values as well as those that are **deficient**. Identify the parameters in your **micronutrient consumption** that are in **excess** of your RDA values as well as those that are **deficient**. Did your **vitamin and mineral average values** meet or exceed the RDAs? How did your **percent calorie distribution** (from seven-day averages) compare to the ideal percent distribution that you chose? How did your fat distribution (see seven-day averages) compare to the recommended 1:1:1 ratio? The minimum length of this discussion is 600 words.

This assignment is worth 50 points maximum.

Checklist

You must have all of the following in this order:

1. Title page (see course policy for details).

2. Computer printouts for seven-day averages. No DAILY foods lists. (Note that you must have these food lists with their nutrients for future reference. DO NOT submit them.)

3. A display of your ideal weight determination, comparison with real weight and the calculations leading to your ideal macronutrient values.

4. A table that compares your average values to your calculated ideals for macronutrients, showing the differences.

5. A table that compares your average values to the RDAs for micronutrients and the differences.

6. A food pyramid showing the comparison between your distribution and the recommended.

7. A discussion containing all of the parts in detail in the discussion section, **labeled** by section and in the order listed.

Note: All course general information rules and regulations regarding papers must be followed.

Quiz

1. The assembly factory for protein in the cell is called

    _____ .

2. The membranous region immediately outside the nucleus is called

    _____ .

3. The internal cell structure which holds the cell's chromosomes is

    _____ .

4. The Golgi complex in the cell wraps cell products in

    _____ .

5. The organelle which has its own DNA and self-replicates is called

    _____ .

6. The organelle which synthesizes complex lipid and carbohydrate is

    _____ .

7. The organelle which makes large amounts of ATP is

    _____ .

8. The small nutrient molecule, glucose, is used in the cell to produce the high energy
    molecule

    _____ .

9. Small nutrient molecules are produced from complex foods by the process of

    _____ .

10. Small nutrient molecules pass into the blood stream via the process called

    _____ .

CHAPTER 2
DIGESTION

Class Notes

Anatomy of the Digestive System

- Mouth—bolus or food ball enters the body
- Esophagus—food tube—connects the mouth to the stomach
- Stomach
- Small Intestine
    - Duodenum
    - Jejunum
    - Ileum
- Large Intestine—colon
    - Caecum (appendix)
    - Ascending Colon
    - Transverse Colon
    - Descending Colon

Digestion

- Carbohydrate
    - Digestion begins in the mouth
    - Pancreatic juice—amylase
    - Digestion continues in the duodenum and jejunum
- Protein
    - Digestion begins in the stomach
    - Pancreatic juice contains proteases
    - Digestions continues in the duodenum and jejunum

- Fats
  - Digestion begins in the duodenum
  - Pancreatic juice contains lipase
  - Digestion continues in the jejunum

Absorption

- Brush border cells absorb most of the nutrients
- Mechanisms of absorption
  - Passive transport—Diffusion
  - Facilitated transport
  - Active transport
  - Endocytosis

Digestion

- Mouth—Ingest (eating), Masticate (chewing).
- Salivary Glands—saliva containing mucus mixes with food, producing slimy, slick, bolus (food ball) so you can swallow.
- Esophagus—a "food tube" passes food from mouth to stomach by rhythmic contractions.
- Cardiac Sphincter—a valve in the stomach that controls the flow of food. Prevents backflow (reflux) from stomach.
- Stomach—Digestive organ in which protein digestion begins. Cardiac sphincter in proximal end of stomach. Pylorus sphincter (distal valve)—will not open until food is digested. It passes food to intestine.
- Gallbladder (Bag)—Stores bile.
- Pancreas (Between stomach + duodenum)—Secretes Pancreatic juice and dumps it in the Duodenum.
- Duodenum (approximately 1 ft. long)—first part of stomach
- Jejunum—where most of digestion and absorption occurs.
- Ileum—distal end of the small intestine. The ileum is clean-up area for absorptions.
- Ileocecal valve (beginning of large intestine)—separates Ileum from Cecum.
- Peristalis—rhythmic contractions of muscles which move the bolus down small intestine and large intestine.
- Any food that is not digested and absorbed goes into large intestine (waste disposal system).

Emulsification of Fat by Bile—Fat Gets Mixed with Water by Emulsification

- **Composition of Bile**
  - Cholesterol

- Bile Salts—synthesized from cholesterol
- Bile Pigments—Bilirubin + Biliverdin
- Lecithin—phospholipid
- Calcium Salts
- **Mucus**—coats lining of stomach to protect stomach
- **Hydrochloric Acid** (the acid of gastric juice)—the H^+ and the Cl^- come from the blood stream.
- **Gastrin**—hormone secreted in the stomach that stimulates secretion of HCl (hydrochloric acid) by the parietal cells and gastric juice by the gastric glands.
- **Histamine**—produced from histadine, causes secretion of mucus; itchy, watery eyes, runny nose, etc., and causes production of hydrochloric acid.
- **Antihistamines**—relieve cold and allergy symptoms.

The stomach produces hydrochloric acid. The acid comes from the bloodstream and leaves bloodstream basic. Hydrogen ion is acid. Sodium bicarbonate (is basic) goes to bloodstream and makes the bloodstream basic. *This is called the Chloride shift.* When the pH in the bloodstream goes up, we get sleepy. Don't give in to sleepiness. We need movement to have good circulation of nutrients. What's happening is blood shifts out of the arms and legs and concentrates around the gut. It is a bad idea to sleep after you eat. In order to digest your food you need energy. If you sleep, everything slows down—the brain activity, metabolism, digestive system, and circulatory system, including the heart rate. We need to move a little, not vigorously.

The parietal cells are located in the base of gastric pit. They take hydrogen ion and chloride ion out of bloodstream and dump it into the lumen of the stomach to create hydrochloric acid (very strong acid). *We produce hydrochloric acid in the stomach to 1) activate the enzymes that are there, and 2) kill bacteria, mold, and fungi in food.*

Influence Hydrochloric Acid Production

- Gastrin—hormone produced by stomach.
- Histamines—produced in response to foreign invaders (bacterial infections, etc.), that create itchy, watery eyes, scratchy throats, secretion of mucus, congestion, sore throats, swollen membranes, swollen bronchial tubes, etc.
- Antihistamines—counteract effects of histamine. Produced by histodine; as adults we make enough; kids do not (controversy if essential amino acid or not). Produces hydrochloric acid.
- Stomach distention—gorging on food produces lots of hydrochloric acid and creates hydrochloric hydria (high hydrochloric acid in stomach).
- Nerve impulses—generated in response to eating and can also be generated in response to stress. When someone is psychologically stressed, it over-stimulates nerve transmission, which produces lots of hydrochloric acid in stomach, causing a burning sensation in stomach. Brain sends nerve signals to region, causing secretion of gastric juice and hydrochloric acid.

- Production of Gastrin—Gastrin is produced in the proximal (beginning) portion of stomach. It is a protein hormone dumped into the bloodstream, travels through bloodstream, and is picked up by receptor sites in the stomach where hydrochloric acid and gastric juice are secreted. So, gastrin stimulates gastric juice and hydrochloric acid secretion.
- Other hormones, other than Gastrin, are CCK and Secretin.
    - Secretin stimulates secretion. It is dumped into the bloodstream, travels up and stimulates secretion of bile to liver and pancreatic juice by the pancreas.
    - CCK—choli (cholesterol) cysto (cyst) Kinin (kenetic), which means action. (Action to the bag of cholesterol or action to the gallbladder). Causes gallbladder to contract and squirt bile out.
    - Secretin causes secretion.
    - CCK causes dumping.
    - Secretin causes production of secretions as they leave organs and go to digestive tract upon the influence of CCK.
    - CCK is the hormone that influences the dumping of pancreatic enzymes.

Components of Pancreatic Juice

- Proteinases—digests protein
- Lipase—digest lipids (more specific—it digests triglycerides)
- Amylase—digests starch
- Bicarbonate—is basic so it neutralizes acid

Hormones of Digestion

Gastrin

- Produced in the stomach

- Stimulates gastric juice secretion

Secretin

- Produced in the duodenum

- Stimulates pancreatic juice secretion

Cholecystokinin

- Produced in the duodenum

- Stimulates the contraction of the gallbladder

Digestion in the Stomach

Enzymes—Proteases

Function

Hydrolyze: Protein ⇨ Smaller Polypeptides

Digestion in the Small Intestine

Pancreatic Enzymes

Proteases: Polypeptides: ⇨ peptides

Amylase: Starch ⇨ Maltose

Lipase: Triglycerides ⇨ Monoglycerides + FA

Digestion in the Jejunum

Intestinal Glands Secrete Intestinal Juice

Peptidase — Peptides $\Rightarrow$ Amino acids

Sucrase — Sucrose $\Rightarrow$ Glucose + Fructose

Maltase — Maltose $\Rightarrow$ Glucose + Glucose

Lactase — Lactose $\Rightarrow$ Glucose + Galactose

Lecithinase

Nucleotidase

Type	Composition	Examples	Physiological Effects
Insoluble Noncarbohydrate	Lignins (woody plant parts)	Wheat and rye bran, grit in fruits, mature vegetables	Antioxidants, binds bile components and minerals
Carbohydrate Cellulose Hemicellulose	Plant cell walls Plant cell well material	Whole grains, some fruits, legumes cruciferous vegetables, Bran of whole grains	Holds water, laxative effect, binds minerals Bulks stool, holds water, binds lipids
Soluble Carbohydrate	Pectins Gums Mucilages Algal substances	Intracellular cement in fruits Cell secretions Cell secretions	Binds lipids, sticky Binds lipids, increases gastric retention Binds bile

Crohn's Disease (Inflammatory Bowel Disease)

- Genetic disorder causing neutrophilic suppression (Antineutrophil cytoplasmic antibodies have been identified).

- Abnormal neutrophil function is well described in Crohn's.

- Mycobacterium avium paratuberculosis has been implicated.

- Environmental factors may lead to colon bacterial population shift, followed by neutrophil suppression.

Quiz

1. Sodium nitrite is a preservative found in

    _____ .

2. Your appetite is controlled by the part of the brain called the

    _____ .

3. The organelle in the cell that makes ATP is called

    _____ .

4. The process of eating food is properly called

    _____ .

5. The conversion of food chunks to small nutrient molecules is called

    _____ .

6. The control center in the cell is called the

    _____ .

7. Genes express themselves by directing the making of

    _____ .

8. The main purpose of catabolism is to produce

    _____ .

9. The synthesis reactions of anabolism results in the formation of the

    _____ .

10. The process by which small nutrient molecules pass across the membranes into the
    blood stream is called

    _____ .

Test 1 Outline

- Absorption of sugars, lipids, amino acids

- Amylase

- Anabolism

- ATP cycle—magnesium

- Basic four food groups

- Basic nutrient classes

- Bile

- Bolus

- Brush border cells

- Carbohydrate digestion

- Catabolism

- Cell membrane composition

- Chylomicrons—lipid micelles

- Chyme

- Coliform bacteria—waste

- Double blind research design

- Duodenum

- Energy components of food vs. fattening

- Enrichment

- Experimental controls

- Fat soluble vitamins

- Fiber—soluble, insoluble

- Food advertising

- Food labels

- Food processing

- Gases and acids in colon

- Genes

- Heartburn

- Hepataic-portal blood system

- Ileum

- Internal organelles membranes

- Jejunum

- Lactase

- Large intestine—colon

- Lipase

- Lipid digestion

- Major minerals

- Maltase

- Mitochondria

- Nucleus

- Nutrient density

- Pancreatic juice

- Protein digestion

- Ribosomes

- Sphincter valve controlled by nerves, hormones

- Sucrase

- Trace minerals

- Types of nutritional research

- U.S. dietary goals

- Villi

- Water soluble vitamins

Test 2 Outline
Class Review Sheet—The Bottom Line

Absorption of Sugars, Lipids, and Amino Acids

You need to know the mechanisms by which we absorb sugars. There are two categories of sugar that we absorb.

- *The G Sugars*—glucose and galactose. Absorbed by active co-transport mechanism along with sodium. Sodium is other player in the act, and then we have an active transport mechanism in which sodium is pumped out of the brush border cells, activating the membrane so the glucose will pass into the bloodstream.
- *Fructose*—the other simple sugar. Fructose absorption is a very slow facilitated transport (protein in membrane that helps us get something across membrane). Fructose is the slowest absorbed of all the sugars.
- *Lipids* are absorbed by micelle absorption and micelle transport. First we form biomicelles and then chylomicron micelles. Lipid (fat) is broken down and surrounded by bile, forming biomicelles. Lipase digests the lipids into fatty acids and monoglycerides which we absorb in the form of micelles. Then once in brush border cells we reassemble them into triglycerides. Then the triglycerides and cholesterol are bound with phospholipids and a protein is added to form chylomicrons. The chylomicrons are then absorbed into the lymphatic system, and we transport the chylomicrons directly into bloodstream bypassing the liver.
- *Amino acids*—amino acids are absorbed by a facilitated transport.

Amylase

- Enzyme produced two places in the digestive tract; in the mouth and pancreas. It is dumped into the duodenum. Digests starch to maltose.

Anabolism

- Metabolic process in which the cells take the small nutrient molecules (SNMs) and assemble them into the compounds of life.
- The compounds of life are protein, complex carbs, complex lipids, and nucleic acid (which is DNA and RNA).

ATP Cycle—Magnesium

- ATP cycle is the cycle in which ATP is used. When its used phosphates are broken off, this forms ADP. ADP is recharged by putting another phosphate back onto it; this requires energy, which we get from some of our SNMs.
- The number one energy nutrient is Glucose. Glucose fuels ATP cycle and magnesium is the coenzyme. So anytime we make ATP we have to have magnesium around.

Basic Four Food Groups

- Meat
- Dairy
- Fruit and vegetable
- Grain

Basic Nutrient Classes

- Protein
- Carbohydrates
- Lipids
- Vitamins
- Minerals
- Water

Bile

- An emulsifying agent
- Produced by liver, stored in gallbladder, made up primarily of cholesterol, bile salts and also contains bile pigments (bilirubin, biliverden).
- Toxic waste the body is trying to get rid of is dumped into the bile and squirted.
- Bile also contains calcium salts and phospholipid Lecithin.
- It's an emulsifying agent for fat and functions in the duodenum to form biomicelles.

Bolus

- A food ball first formed in the mouth goes through esophagus to stomach where it becomes chyme.
- Chyme is the acidic food ball.

Brush Border Cells

- Cells that line the villi (humps in the small intestinal tract), have microvilli on the edge of them; that's why they are called brush.

Carbohydrate Digestion

- Complex carbs are called starch; starch is digested by amylase into maltose.
- Maltose is digested by maltase into glucose (2 glucose molecules).
- Sucrase digests sucrose into lactase that digests lactose into

Catabolism

- Metabolic process of making ATP; occurs primarily in the mitochondria.
- The main fuel for catabolism is glucose.

Cell Membrane Composition

- Cell membranes are made of phospholipids.

Chylomicrons

- Lipid micelles. They are lipoproteins, so they contain lipid in the middle (i.e. cholesterol and triglycerides) surrounded by phospholipids with a little protein (B48 protein) imbedded in the surface. B48 has specific shape that fits B48 receptor sites.

Chyme

- Acidic food ball produced in the stomach that has to be neutralized in duodenum by pancreatic secretion and bile (both are basic).
- Bicarbonate in pancreatic juice neutralizes the chyme.

Coliform Bacteria

- Bacteria in colon.
- Produce acids and gases when it partially digests fiber.

Double-Blind Research Design

- Participants and researchers are both blind as to who is getting what (placebos).
- No one directly involved in the clinical trials knows what is going on.

Duodenum

- First segment of small intestine
- Mixing area for pancreatic juice, chyme, and bile.

Kcal Components

- Carbs, proteins and fats.
- Each kcal nutrient has a certain number of kcals per gram.
- Carbs and proteins have 4 kcal/g and fat has 9 kcals/g.

Enrichment

- Nutrients put back in food after processing.
- Controlled by law.
- Put back B-1 Thiamin, B-2 Riboflavin, B-3 Niacin, B-9 Folacin, and the mineral Iron.

Experimental Control

- People participating in a piece of research who are given a fake (placebo).
- The best experimental design is the double-blind placebo controlled trial.

Fat Soluble Vitamins

- A, D, E, and K
- They dissolve in fat, not in water.

Fiber-Soluble and Insoluble

- Carbohydrate material that is not digestible.
- Insoluble are very large—too big to be suspended in water. So big you can see the with naked eye.
- Insoluble fiber promotes rapid transit (gets rid of waste fast).
- Soluble fiber bonds to fat small enough to be suspended in water; it disappears.

Food Advertising

- No matter how much food advertising you see it will not change the amount of food you eat.
- The amount of food you eat is controlled by your hypothalamus in the brain.
- Advertising shifts food choices, but not the amount.

Food Label

- Tells you serving size, how many servings in package, composition in terms of carbs, protein fat, fiber, and vitamin and mineral content.

Food Processing

- Generates foods that are decreased significantly in nutrients.
- Most nutrients are removed.
- They give back some in enrichment.
- Main reason we process food is to increase shelf life of food.

Gases and Acids

- In colon.
- Produced by bacteria as they act on plant fibers.

Genes

- Pieces of DNA found in the nucleus of the cell.

Heartburn

- The stomach acid splashing back up on the esophagus, burning the tissue of the esophagus.

CHAPTER 3
CARBOHYDRATES

Class Notes

Chemical Structure of Monosaccharides—
Their Dietary Sources and Physiological Significance
Artificial Sweeteners

- Trioses (Aldoses vs. Ketoses)
    - Glyceraldehyde
    - Dihydroxyacetone
- Pentoses
    - Ribose
    - Deoxyribose
- Hexoses
    - Glucose
    - Fructose
    - Galactose

Chemical Structure of Disaccharides

- Maltose
- Sucrose
- Lactose

Chemical Structure of Oligo and Polysaccharides (Polymers)

- Starch (amylose vs. amylopectin structure)
- Glycogen
- Cellulose

The Physiology of Glucose

- Absorption of carbohydrates
  - Liver processing of monosaccharides
- Pancreatic endocrine secretions
  - Insulin—production and function
    - Cephalic phase
    - Pancreatic phase—beta cells
- Functions
  - Insulin receptors
  - Lipogenesis
  - Glycogenesis
  - Proteogenesis
- Glucagon—production and function
    - Production—Alpha cells
- Functions
  - Glycogenolysis
  - Proteolysis
  - Lipolysis
- Glycemic States—Glucose tolerance curves
  - Normoglycemia
  - Hypoglycemia—causes, symptoms and treatment
  - Hyperglycemia—causes, symptoms and treatment
  - Gluconeogenesis—synthesis of denovo glucose
    - Muscle wasting—damage
    - Ketosis—a metabolic hazard
    - Nitrogen waste increase

Glycemic Index—Guide for Selection of Best Carbo Sources

- Application to nutritional carbohydrate choices.
- Application in the prevention and treatment of abnormal glycemic states.

Dietary Fiber

- Sources—various types
- Functions
  - Peristalsis
  - Lipid (cholesterol) absorption
  - Colon disease

Monosaccharides

Glucose

- Blood of Grape sugar

- Must maintain blood level (Fasting: 70–100 mg/dl)

- Rapidly absorbed

- Sweetness < sucrose

- Liver converts all sugars to glucose

Fructose

- Fruit sugar (strawberries, plums, apples, peaches)

- Sweetest sugar

- Slowest absorbed sugar (facilitated transport)

Galactose

- Not found free in nature

- Component of milk sugar (Lactose)

- Necessary in synthesis of neurocellular complexes

- Absorbed like glucose

- Not very sweet

Must Recognize The "G" Sugars

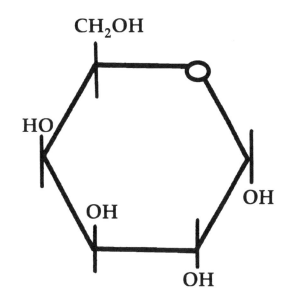

Galactose

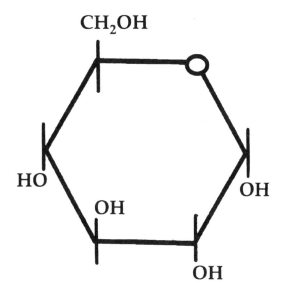

Glucose

G looks like 6

Six carbon sugars
Six membered ring

The "F" Sugar

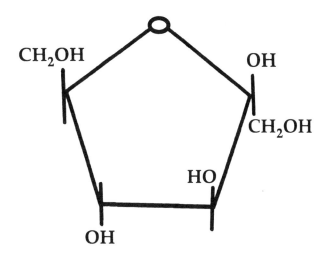

"F" Five Membered Ring

Fructose

Disaccharides

Maltose

- Malt Sugar—geminated seeds

- Glucose—Glucose

- Starch → Maltose → 2 glucose

- Malt confectionaries and brewed beverages

Sucrose

- Cane or beet sugar (table sugar)

- Standard for sweetness

- Glucose—Fructose

Lactose

- Milk sugar

- Glucose—Galactose

- Lactose intolerance

Polysaccharides

Glycogen

- Polymer of glucose—highly branched

- Made in the liver of animals

- Can be made and stored in muscle (need conditioning)

- Animal liver is the only food source (environmentally undesirable due to man-made toxins)

Starch

- Plant energy storage product (seeds, roots, stems)

- Digests quickly to maltose, then to glucose (see glycemic index)

- Complex carbohydrate

Fiber
Soluble

- Found in fruits, vegetables, oats, barley, beans

- Attracts water—bulks the stool

- Binds to cholesterol and fats, thus inhibits absorption

- Pectins, gums, mucilages, lignins (non-carbohydrate)

- Slows gastric emptying

Insoluble

- Found in whole grains (wheat, rice, corn, rye)

- Encourages peristalsis

- Laxative action

- Holds water

- Binds minerals

Blood Glucose

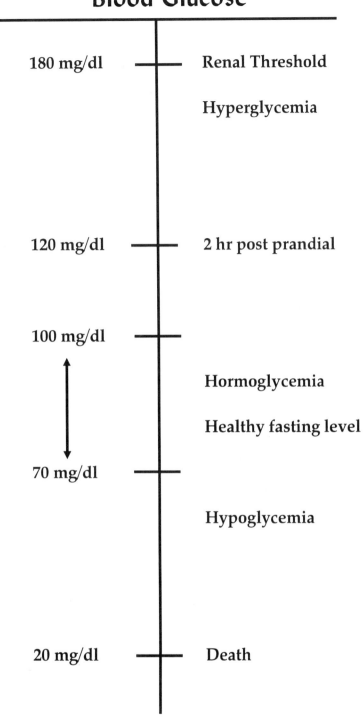

180 mg/dl	—	Renal Threshold
		Hyperglycemia
120 mg/dl	—	2 hr post prandial
100 mg/dl	—	
	↕	Hormoglycemia
		Healthy fasting level
70 mg/dl	—	
		Hypoglycemia
20 mg/dl	—	Death

Comparison Chart of Fiber Sources

Food (100g)	Energy (kcals)	Carbo (g)	Fat (g)	Fiber (g)	ND fiber
Apples	59	15	0.5	2.8	0.05
Avocado	112	8.9	8.9	5.3	0.05
Blackberries	52	12.5	0.7	5.6	0.10
Figs, dried	255	65	1	9.5	0.04
Prunes, cooked	107	28	0.4	6.5	0.06
Rasberries	49	11.4	0.8	6.5	0.13
Mango	65	17	0.5	2	0.03
Bread, multigrain	250	46	4	7.7	0.03
Barley, cooked	123	28	0.6	3.8	0.03
Cereal, All Bran	258	74	3.2	32	0.13
Granola, Nature Valley	451	65	17.7	6	0.01
Cereal, Fruit 'n Fiber	339	75	5.3	14	0.04
Rice, brown	111	23	1	2	0.02
Almonds	589	20	52	10.6	0.02
Artichoke	45	9	0.6	4.8	0.11
Black beans	132	23	0.23	8.2	0.06
Broccoli spears	28	5	0.6	2.8	0.10
Sundried tomatoes	257	56	3.7	13	0.05

Low Glycemic Index Foods

All-bran	60	Lima beans, baby, frozen	46
Apple juice	58	Milk + 30 g. bran	38
Apple	54	Milk + custard + starch + sugar	61
Apricots	44	Milk, chocolate, artificially sweet	34
Barley chapatti	61	Milk chocolate, sugar sweetened	49
Barley	49	Milk, full fat	39
Beans, dried, not specified	40	Milk, skim	46
Black beans	43	Peach, fresh	60
Black gram	61	Peanuts	21
Black-eyed beans	59	Pear, fresh	53
Brown beans	34	Peas, dried	32
Brown beans	54	Pinto beans	55
Bush honey, sugar bag	61	Plum	55
Butter beans + 10 g. sucrose	44	Ravioli, durum, meat filled	56
Butter beans + 5 g. sucrose	43	Rice bran	27
Butter beans	44	Rye	48
Cheeky yam	49	Sausages	40
Cherries	32	Soya beans	25
Chick peas (garbanzo beans)	47	Soya beans, canned	20
Chick peas, canned	60	Spaghetti, boiled 5 min	52
Chick peas, curry, canned	58	Spaghetti, protein enriched	38
Corn hominy	57	Spaghetti, white	59
Fettuccine	46	Spaghetti, wholemeal	53
Fructose	32	Spirali, durum	61
Grapefruit	36	Split peas, yellow, boiled	45
Green gram	54	Star pastina	54
Haricot/navy beans	54	Tomato Soup	54
Kidney beans	42	Tortilla	
Kidney beans, autoclaved	49	Vermicelli	50
Lentils, green	42	Wheat kernels	49
Lentils, not specified	41	Yogurt, low fat, artificially sweet	20
Lentils, red	36	Yogurt, low fat, fruit sugar sweet	47
Lima beans broth	51	Yogurt, unspecified	51

Moderate Glycemic Index Foods

Baked beans, canned	69	Macaroni	64
Barley kernel bread	66	Mixed grain bread	69
Bran Buds	75	Oat bran bread	68
Bread (Acacia coriacea)	66	Orange juice	74
Bulger bread	75	Orange	63
Cake, banana, made with sugar	67	Peach, canned	67
Cake, sponge	66	Pear, canned	63
Capellini	64	Peas, green	68
Chocolate	70	Pineapple juice	66
Fruit loaf	67	Pinto beans, canned	64
Grapefruit juice	69	Pumpernickel	71
Grapes	66	Red River Cereal	70
Ice cream, low fat	71	Rice, instant, boiled 1 min	65
Instant noodles	67	Rice, parboiled	68
Kidney beans, canned	74	Rice, parboiled, high amylose	69
Kiwifruit	75	Romano beans	65
Lactose	65	Rye Kernel bread	66
Lentil soup, canned	63	Sweet potato (Ipamoea batatas)	63
Lentils, green, canned	74	Tortellini, cheese	71
Linguine	65	Yam	73

High Glycemic Index Foods

Minimize your intake (always include a low GI food)

Arrowroot	95	Honey	104
Bagel, white	103	Ice cream	87
Banana	77	Jelly beans	114
Banana, unripe, steamed	100	Kaiser rolls	104
Barley flour bread	95	Life Savers	100
Beans, dried, P. vulgar	100	Life	94
Beets	91	Linseed rye bread	78
Black bean soup	92	Lucozade	136
Bran Chex	83	Macaroni and Cheese	92
Bread stuffing	106	Maize chapati	89
Breadfruit	97	Maize meal porridge, refined	106
Breakfast bar	109	Maize meal porridge, unrefined	101
Breton Wheat Cracked	96	Maize	98
Broad beans (fava beans)	113	Maltose	150
Buckwheat	78	Mango	80
Butter beans + 15 g. sucrose	77	Mars Bar	97
Cake, angel food	95	Melba toast	100
Cake, banana, w/o sugar	79	M'fino wild greens	97
Cake, flan	93	Millet	101
Cake, pound	77	Morning Coffee cookies	130
Carrots	101	Muesli Bars	87
Cheerios	106	Muesli	80
Cocopops	110	Muffins	88
Cordial, orange	94	Nutri-grain	94
Corn Bran	107	Oat Bran	78
Corn Chex	118	Oat kernel bread	93
Corn chips	105	Oatmeal	79
Cornflakes	119	Parsnips	139
Couscous	93	Pastry	84
Cream of Wheat	100	Pawpaw	83
Crispix	124	Pineapple	94
Croissant	96	Pita bread, white	82
Crumpet	98	Pizza, cheese	86
Donut	108	Popcorn	79
French baguette	136	Porridge	87
French fries	107	Post Flakes	114
Fruit cocktail	79	Potato crisps	77
Glucose tablets	146	Potato mashed	100
Glucose	137	Potato, baked	121
Gnocchi	95	Potato, instant	118
Golden Grahams	102	Potato, microwaved	117
Graham Wafers	106	Potato, new	81
Grapenuts	96	Potato, Pontiac, boiled	80
Green pea soup, canned	94	Potato, Prince Edward Island	87
Hamburger bun	87	Potato, steamed	93
High Fibre Rye Crispread	93	Potato, white, not specified	80

Glucose Transport

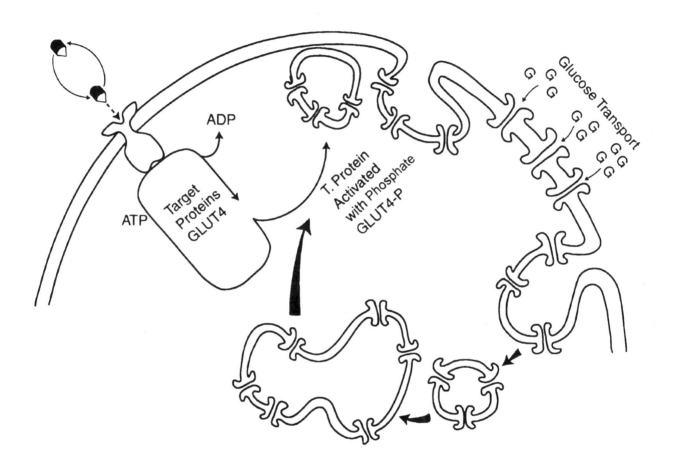

Blood Glucose vs. Time

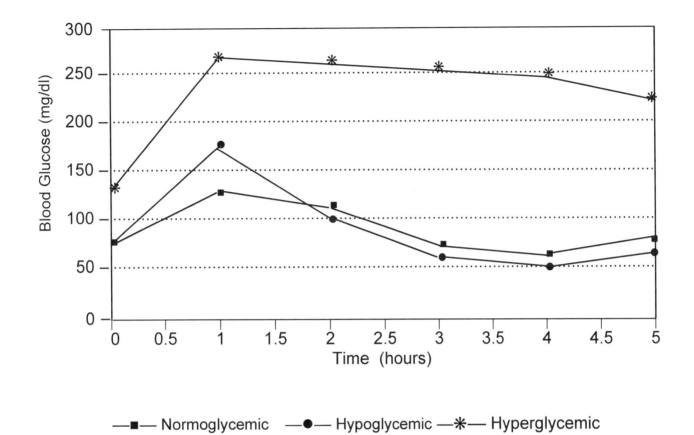

Symptoms of Hypoglycemia

- Headache

- Lethargy—fatigue, weakness, sleepiness

- Irritability—inappropriately

- Mental Confusion—impaired ability to concentrate and learn

- Shaky—trembling

- Anxiety—tense, miserable, fearful

- Diaphoretic

- Depression

More Severe Symptoms

- Nausea

- Jekyll and Hyde Behavior

- Dizziness

- Impaired Vision

- Faint/Coma—usually drug induced

- Death

Insulin Independent Organs

Liver

Brain (CNS)

Kidney

Types of Diabetes

Type I

Insulin Dependent Diabetes Mellitus (IDDM)

Type II

Non-Insulin Dependent Diabetes Mellitus (NIDDM)

Gestational Diabetes (Hormonally Induced)

Diabetes Insipidus (Pituitary Disease—Vasopressin Deficiency)

Symptoms of Diabetes

IDDM
Insulin Dependent Diabetes

- Type I
- Juvenile onset
- 20% of diabetics

1. Insulin deficiency

2. Excessive hunger (early stages)

3. Excessive thirst (kidneys excrete to lower blood sugar)

4. Frequent urination

5. Decreased body mass (metabolic shift to lipid catabolism)

6. Slow wound healing

7. Frequent illness

NIDDM
Non-Insulin Dependent Diabetes Mellitus

- Type II
- Adult onset
- 80% of diabetics

1. Excessive hunger → Overnutrition

2. Obesity (liver uptake of glucose is independent of insulin) glucose → fat

3. Normal to excessive insulin levels

4. Excessive thirst/urination

5. Sedentary lifestyle—inactivity

Complications of Diabetes

1. Atherosclerosis

2. Impaired Circulation (blood clotting)

3. Eye Pathology (eventual blindness)

4. Kidney Failure

5. Cerebrovascular Accident (CVA—Stroke)

6. Cardiac Arrest

7. Immune Suppression

Quiz

1. List two simple sugars in lactose

    _____ .

2. Name the hormone that initiates glucose transport

    _____ .

3. The hormone that releases glucose from the liver is

    _____ .

4. Name the simple sugars that are found in sucrose

    _____ .

5. High blood sugar is called

    _____ .

6. Name two kinds of dietary fiber

    _____ .

7. The carbohydrate polymer that is made by plants for energy storage is

    _____ .

8. Insulin initiates glucose transport when it is received by the

    _____ .

Quiz

1. Fiber which slows glucose absorption and binds to cholesterol is

    _____.

2. Fiber which increases peristalsis and decreases transit time is

    _____.

3. Lactose intolerance is a decrease in the production of the enzyme

    _____.

4. The higher your blood sugar arises after eating, the higher the blood level of the hormone

    _____.

5. The organs in the body which receive glucose independently of insulin include: brain, kidneys and

    _____.

6. Complex carbohydrates which have been processed have had the germ removed and also the

    _____.

7. When blood sugar exceeds the renal threshold, glucose abnormally appears in the

    _____.

8. Normoglycemia implies a fasting blood glucose level between

    _____.

9. Neurotransmitters ultimately control blood glucose levels; they are produced in the

    _____.

10. The No. 1 artificial sweetener used in the U.S. today is

    _____.

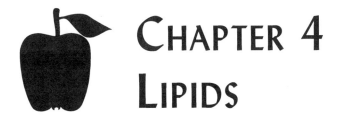

CHAPTER 4
LIPIDS

Class Notes

Fatty Acids

- Saturated
- Monounsaturated
- Polyunsaturated (PUFAs)—Essential Fatty Acids (EFAs)
  - Omega 3
  - Omega 6

Triglycerides

- Glycerol
- 3 fatty acids

Phospholipids

- Diglyceride
- Phosphate
- Alcohol (choline, inositol, serine, ethanolamine)

Cholesterol and Derivative Sterols

Dietary Sources of Various Fatty Acids

Hydrogenation of Fatty Acids—The Trans-Fatty Acid Controversy

Cell Membrane Structure

- Phospholipid bilayer
- Cholesterol
- Surface proteins
- Transmembranic proteins

- Internal proteins
- Oligosaccharides

Lipid Micelles—Absorption

Lipoproteins

- Chylomicrons
- VLDL
- LDL
- HDL

Lipid Oxidation—Formation of Malondialdehyde

Need for Antioxidants

Cholesterol—Cell Membranes

- Bile salts
- Steroid hormones
- Vitamin D
- Anabolic Steroids—man-made

Prostaglandins

- Made from PUFAs
- Regulate cellular function

Comparison Chart of Fat Sources (100 g Each)

Food	Energy	T. Fat	S. Fat	M-USF	P-USF	S: M: P Ratio	Chol
Egg, whole	148	10	3	3.8	1.4	2.1: 2.7: 1	426
Butter	716	81	51	24	3	17: 9: 1	219
Canola Oil	884	100	7	59	30	1: 8.4: 4.3	0
Olive Oil	884	100	13.6	74	10	1.4: 7.4: 1	0
Avocado (FL)	112	9	1.7	4.9	1.5	1.1: 3.3: 1	0
Salmon (broiled)	216	11	1.9	5.3	2.4	1: 2.8: 1.3	87
Tuna (bluefin fresh)	188	5.3	1.2	1.6	1.4	1: 1.31.2	38
Pork chop	239	13.5	5	6	1.1	4.5: 5.6: 1	80
Beef prime rib	376	31	14	15.8	1.3	11: 12: 01	85
Peanuts	581	49	6.8	24.5	19	1: 3.6: 2.8	0
Almonds	589	52	4.9	34	11	1: 7: 2.2	0
Cashews	574	47	9.3	27	7.8	1: 1.2: 3.5	0
Walnuts	607	57	3.8	12.7	37.5	1: 2.6: 9.8	0
Pecans	667	68	5.4	42	17	1: 7.8: 3.1	0

Oxidative Tissue Damage and Antioxidants

Oxidative Species

- The oxygen molecule in the air.

- Free radicals from air pollution.

  - NO^- (nitric oxide) from automobile exhaust, also produced by macrophages.

  - O_2- (superoxide) from mitochondrial metabolism.

  - H_2O_2 (hydrogen peroxide) from cellular metabolism. Not a free radical, but produces free radicals.

  - OH^- (hydroxyl) produced when NO reacts with O_2, then decomposes in the presence of H^+.

- Free radicals are produced in large numbers when excess unbound iron or copper are present.

- A barrage of free radicals is produced during periods of immune excitation, i.e., inflammation, infection, stress-induced alarm state, generation of epinephrine.

Oxidative Tissue Damage

- Oxidation of circulating lipids, proteins, etc.

- Oxidation of external cell membrane PUFAs with subsequent generation of cell membrane "holes" and cytotoxic aldehydes.

- Oxidation of internal organelle membranes (i.e., endoplasmic reticulm, mitochondria, nucleus).

- Oxidation of membrane proteins, both internal and external.

- Oxidation of DNA resulting in gene mutation and subsequent production of aberrant protein.

Protection From Lipid Oxidation

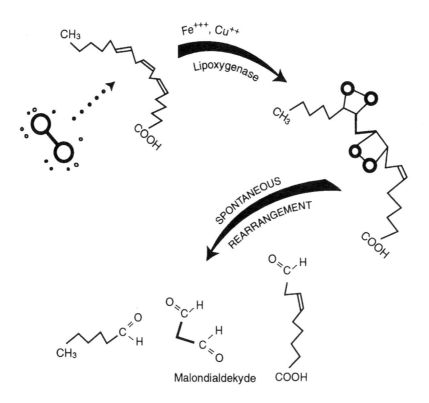

Vitamin/Mineral Protection

Ascorbic acid (vitamin C)	1000—2000 mg/day
Beta carotene	25,000 IU/day
D-Alphatocopherol (vitamin E)	400 IU/day
Selenium	400 ug/day
Zinc	35 mg/day
Manganese	26 mg/day

Enzyme Antioxidant Systems

Glutathione Peroxidase Se = coenzyme

$$O_{2=} + 2\,GSH \rightarrow GS - SG + HOH$$

Superoxide Dismutase

Mitochondrial SOD	Zn & Mn coenzymes
Cytoplasmic SOD	Zn & Cu coenzymes

$$O_{2-} + 2H \rightarrow H_2H_2 + O_2$$

Catalase   Zn = coenzyme

Destroys hydrogen peroxide

EFA Anabolism (Prostaglandin Formation)

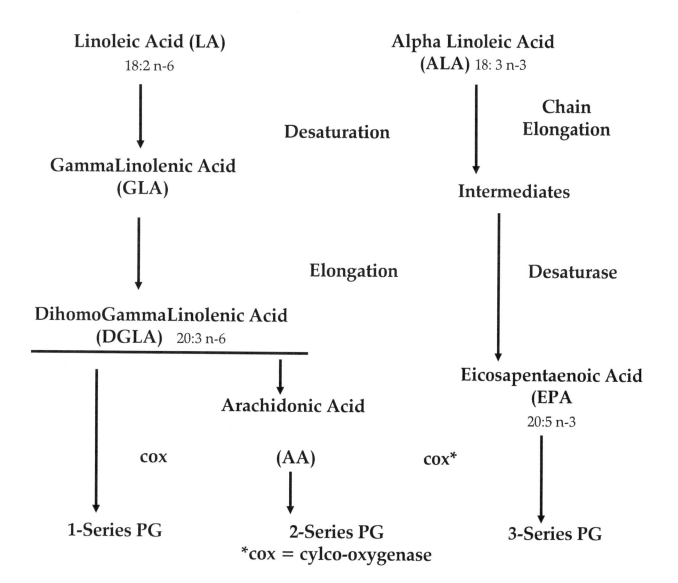

Linoleic Acid (LA)
18:2 n-6

Alpha Linoleic Acid (ALA) 18: 3 n-3

Desaturation

Chain Elongation

GammaLinolenic Acid (GLA)

Intermediates

Elongation

Desaturase

DihomoGammaLinolenic Acid (DGLA) 20:3 n-6

Eicosapentaenoic Acid (EPA 20:5 n-3

Arachidonic Acid

cox

(AA)

cox*

1-Series PG

2-Series PG
*cox = cylco-oxygenase

3-Series PG

Quiz

1. List the components of a *triglyceride*.

    _____

    _____

    _____

    _____

2. Cholesterol is made in the body from

    _____ .

3. The number of fatty acids in a diglyceride is

    _____ .

4. The number of hydrogens which are missing in a monounsaturated fatty acid is

    _____

5. List the alpha-numeric designations for the two EFAs.

    _____

    _____

    _____

6. List the names of the two EFAs.

    _____

    _____

    _____

Test 1 Outline

- Absorption of lipids

- Carbohydrate foods

- Chemical composition of triglycerides

- Common sweeteners used in U.S.

- Complex carbohydrate

- Composition of canola oil vs. olive oil

- Composition of lecithin and phospholipids

- Condensation synthesis

- Dental caries and carbohydrate

- Dextrose

- Diabetes—characteristics

- Dietary sources of saturated fat

- Functions of fat

- Gluconeogenesis

- Glycemic Index

- Glycemic response

- High GI food vs. low GI foods

- Hormones controlling glycemic response

- Hydrogenation

- Importance of soluble and insoluble fiber

- Lactose intolerance

- Lipoprotein composition

- Membrane fluidity and PUFAs

- Need for carbohydrates in the diet

- Olesta

- Omega 3 EFA—alphalinolenic acid—ALA – 18:3 n-3

- Omega 6 EFA—linoleic acid—LA 18:2 n-6

- Partial hydrogenation

- Problems getting rid of cholesterol

- Sources of cholesterol

- Sources of EFAs

- Sources of fiber

- Sources of simple sugars

- Stress hormone—action on glucose

- Structures of simple carbohydrates

- Structures of unsaturated fatty acids

- Sweetness of simple sugars

- Synthesis and storage of glycogen

- Synthesis of lipoproteins—tissues

- U.S. recommendation for fat

- Uses of phospholipids and cholesterol

Test 2 Outline
Class Review

Absorption of Lipids

- Lipids need help to get absorbed.
- Bile is an emulsifying agent for fat.
- We absorb them into brush border cells, then the brush border cells package them into chylomicrons.
- Chylomicrons are then smuggled through the lymphatic system into the blood stream.

Carbohydrate Foods

- Most fruits and vegetables, nuts, seeds, grains, legumes (plant reproductive structures)

Chemical Composition of Triglycerides

- Made of a glycerol and 3 fatty acids

Common Sweeteners in U.S.

- Sucrose
- Nutra-sweet (aspartame)

Complex Carbohydrates

- Starches
- Nuts, grains, seeds, legumes
- Some vegetables/potatoes, rice, pasta

Composition of Canola Oil vs. Olive Oil

- Canola
  - High in monounsaturated
  - 2.3/1 ratio
  - Omega 6 / omega 3
  - Lowest saturated fat
- Olive
  - High in monounsaturated
  - Low saturated fat

Composition of Lecithin and Phospholipids

- Phospholipids
  - Composed of diglycerides (glycerol and 2 Fas)
  - Plus, on the third "C" on the glycerol we hang a phosphate
- Lecithin
  - Alcohol bonded onto a phosphate in addition (to a phospholipid), which makes it a specific phospholipid. These phospholipids are the primary component of cell membrane structure, stabilized by cholesterol.

Condensation Synthesis

- Synthesis by the removal of water
- This is the way we put together all biological molecules.
- Ex: when we put fatty acids on a glycerol; when we put glucose together to form chains

Dental Caries

- Only thing we know for sure that carbs cause
- From a metabolic point of view, could lead to over production of fat and saturated fat production leads to synthesis of cholesterol

Dextrose

- Another name for glucose

Diabetes

- Type I
  - Insulin dependent
  - Beta cells of pancreas damaged, don't produce insulin.
  - If you don't produce insulin you don't turn on glucose transport system and the glucose stays in the blood; glucose level goes higher and higher.
  - People with Type I Diabetes have a lean body frame.
  - They burn fat and protein so they don't grow very well.
- Type II
  - NIDDM—Normal Insulin
  - Obese, sedentary
  - Some problem with insulin receptor site
  - Are being damaged or destroyed so we are not getting insulin and we don't turn on glucose transport system

- Three Tissues that get glucose, independent of Insulin
    - Brain
    - Liver
    - Kidneys

Dietary Sources of Saturated Fat

- Animal sources
- Coconut and Palm Oil

Functions of Fat

- We use fat to make cell membrane more than anything
- Fat deposits in adipose tissue which insulates body from loss of heat
- Cushions organs
- Unsaturated fat used to make prostoplandins—internal cell regulators

Gluconeogenesis

- Process of making new glucose
- In response to no carbohydrate intake
- Message goes from brain to liver—"Make more glucose. I can't live without it."
- We make glucose from amino acids, we get amino acids from dietary protein or breaking down of muscle tissue.
- We need to eat carbohydrates to avoid gluconeogenesis, because it almost always involves breaking down of muscle tissue.

Glycemic Index

- Used to judge quality of carbs
- Carbs absorbed quickly cause a big glucose spike in bloodstream and big insulin response.
- Carbs absorbed slowly, low glucose response, low insulin response. Most desirable.

Glycemic Response

- When you eat a carbohydrate you release insulin.
    - A proper glycemic response is to release a little bit of insulin.
    - An abnormal response is to eat a carbohydrate and it produces a large amount of insulin.
    - When you produce a lot of insulin you have a rapid glucose transport and your blood sugar is going to fall rapidly and drastically.

Insulin

- Protein hormone produced in pancreas (beta cells)
- Turns on glucose transport mechanism
- Causes body to pump glucose out of the bloodstream and into the cells, causing blood sugar to fall

Glucagon

- Protein hormone produced in alpha cells of pancreas
- Promotes liver to dump its stored glycogen
- Muscle glycogen can only be used in-house

Hydrogenation

- Turns unsaturated fat into saturated fat

Soluble Fiber

- Fruits and vegetables
- Bonds to cholesterol and lipids; bile helps body get rid of them
- Lack of soluble fiber can lead to reabsorption, and elevated levels in bloodstream

Insoluble Fiber

- From grains
- Promotes rapid transit

Lactose Intolerance

- Lack of lactase production
- Can't break up lactose properly
- Goes to colon, colon bacteria ferment it into CO_2 and acid.

Lipoprotein Composition

- Chylo-dietary lipop made mostly of triglycerides, protein, phospho, cholesterol sent to liver; liver breaks up chylo and reconstructs them into VLDL
- VLDL primarily transports triglycerides
- LDL primarily transports chol to cells
- HDL transports chol back to liver from cells

PUFAs

- Polyunsaturated fatty acids—cause membrane fluidity

Need for Carbs in Diet

- Brain must have carbs. If you don't eat them you spend muscle protein to make glucose

Olestra (sucrose polyester)

- Fake fat
- Sucrose with FS hanging off of it
- Can't digest it
- Bonds to fat soluble vitamins so you can't absorb them

Omega 3 EFA

- Alphalinolenic Acid
- ALA
- 18: 3 n-3

Omega 6 EFA

- Linoleic Acid
- LA
- 18: 2 n-6

Partial Hydrogenation

- Limiting breaking of double bond
- Changes from cis to trans
- Promotes heart disease

Problems Getting Rid of Cholesterol

- Bile. We dump it as bile; body can't get rid of it any other way

Sources of Cholesterol

- Animal tissues

Source of EFAs

- Omega 3-flaxseed
- Omega 6-vegetable oil

Sources of Fiber

- Soluble—fruits and vegetables
- Insoluble—whole grains

Sources of Simple Sugars

- Table sugar mono and disaccharides—honey
- Milk-lactose, fruit—fructose

Stress Hormone—Action on Glucose

- Epinephrine—causes glucose to be released from liver
- Causes liver to dump its glycogen stores

Structure of Simple Carbs

CHAPTER 5
PROTEINS

Class Notes

Protein Structure

- Made up of amino acids, held together by peptide bonds, side chain dictates three-dimensional shape; dimensional shape.

N Balance N in = out

- Know the positive and negative balance. We strive for N in equaling n out. Ex: a positive N balance would mean that we are growing something like a fetus, making new tissues or making muscle protein.

Enzymes

- Catalyze chemical reactions. Ex: proteins work because of their three-dimensional shapes. Anything that disrupts their three-dimensional shapes disrupts their function.

Antibodies

- Immune proteins that neutralize invader molecules in the body.

Hormones

- Chemical messengers. Ex: Insulin

Receptor Sites

- Proteins in cell membranes that receive specific messages from other hormones or signaling molecules.

Transporters

- Molecules that carry. Ex: hemoglobin carries oxygen. LDL, HDL, VLDL

Proteins

Chemical Structure

- Amino Acids

- Peptides—oligos

- Proteins

Digestion and Absorption

- Digestion of proteins and peptides

- Absorption of Amino acids, di and tripeptides

Protein Metabolism

- Anabolism of amino acids

- Catabolism of protein—amino acids

Protein Functions

- Enzymes

- Antibodies, etc.

Proteins in Food

- Quality

- Nutrient Density

- Efficiency

Health Effects

- Maramus

- Kwashiorkor

Amino Acids in Protein

20 Different Amino Acids

8/9 Essential Amino Acids

Amino Acids

- 9 Essential

- 11 Non-Essential

Classification

- Acidic Amino Acids

- Basic Amino Acids

- Neutral (Hydrophobic) Amino Acids

- Sulfur Amino Acids

- Polar Amino Acids

Essential Amino Acids

- Histidine

- Isoleucine

- Leucine

- Lysine

- Methionine

- Phenylalanine

- Threonine

- Tryptophan

- Valine

Non-Essential Amino Acids

- Alanine

- Arginine

- Asparagine

- Aspartic Acid

- Cystine

- Glutamic Acid

- Glutamine

- Glycine

- Proline

- Serine

Non-Protein Amino Acids

Carnitine

Gamma Aminobutyric Acid

Taurine

Amino Acid Deficiencies in Plant Sources of Protein

Plant Source	Amino Acid Deficiencies
Corn	Tryptophan, Threonine
Cereal Grains	Lysine
Legumes	Methionine, Tryptophan
Peanuts	Mithionine, Lysine
Rice	Tryptophan, Threonine
Soybeans, Black Beans	Methionine

Protein Efficiency

Food Source	Percent Incorporation	Percent Waste
Eggs	94	6
Milk	82	18
Fish (average seafood)	80	20
Cottage Cheese	75	25
Cheese	70	30
Rice (whole grain)	70	30
Red Meat and Poultry	67	33
Soybeans, Black Beans	61	39
Wheat (whole)	60	40
Cashews	58	42
Lima Beans	52	48
Corn	51	49
Walnuts	50	50
Peas	47	53
Peanuts	43	57
Kidney Beans	38	62
Lentils	30	70

Comparison Chart of Protein Foods

Food	Amount	Energy (kcal)	Protein (g)	Carbo (g)	Fat (g)	Chol (mg)
B. rice	0.5 cup	108	9% 2.4	83% 22.4	7% 0.8	0
Blk beans	0.5 cup	113.5	26% 7.4	70% 20	4% 0.5	0
Combined	1 cup	221.5	17% 9.8	77% 42.4	5% 1.3	0
Hamburger	3 oz	231	37% 21.4		63% 16	86
Beef p.rib	3 oz	319	24% 19		73.3% 26	72
Egg poach	3 oz	126	34% 6.2	3%	62% 8.5	212
Ham, lean	3 oz	123	58.5% 18	1.5	33% 4.5	45
Chicken w	3 oz	145	71.7% 26		7% 1.2	75
Chicken d	3 oz	173	53% 23		42% 8	78
Turkey w	3 oz	133	78% 26		7% 1	59
Milk whole	3 oz	51	19.6% 2.5	29% 3.7	44% 2.5	11
Milk skim	3 oz	29	34.5% 2.5	52.4% 3.8	0% 0	1.3
Grouper	3 oz	100	89% 22		11% 1.2	40
Lobster	3 oz	100	89% 22		6% 0.2	60
Salmon	3 oz	157	62% 24		38% 6.6	69
Shrimp	3 oz	84	90% 19		10% 0.9	168

Rice Comparison Table (0.5 Cup)

Nutrients	Brown Rice	White Rice	Wild Rice
Kcals	110	133	83
Protein, g	2	1	3.5
Carbohydrates, g	23	25	18
Fiber, g	1.7	0.2	1
Total Fat, g	1	0	0.5
Calcium, mg	10	3	3
Sodium, mg	0	0	3
Potassium, mg	77	30	83
Zinc, mg	0.6	0.4	1.1
Iron, mg	0.5	1.6	0.5
Magnesium, mg	43	14	26
Vitamin A, RE	0	0	0
Vitamin C, mg	0	0	0
Thiamin, mg	0.1	0.17	0.04
Riboflavin, mg	0.01	0.01	0.07
Niacin, mg	1.3	1.9	1.1
Vitamin B6, mg	0.15	0.05	0.11
Folacin, ug	4	2	22
Vitamin B12, ug	0	0	0
ND wrt protein	0.02	0.01	0.04

The Egg

Parameter	Yolk (17g)	White (33g)	Ratio
Calories	63	16	4:1
Protein g	2.8	3.3	1:1.2
Fat g	5.6	1	
Saturated Fat g	1.7	0	
Unsaturated Fat g	3.0	0	
Cholesterol mg	272	0	
Carbohydrate g	0.04	.41	
Calcium mg	26	4	6.5:1
Iron mg	0.95	.01	95:1
Magnesium mg	3	3	1:1
Manganese mg	.015	.013	1:1
Potassium mg	15	45	1:3
Selenium ug	2.96	1.88	1.6:1
Sodium mg	8	50	1.6:3
Zinc mg	0.58	0.01	58:1
Vitamin A (RE)	62.1	0	
Thiamin mg	0.043	0.002	21.4:1
Riboflavin mg	0.074	0.094	1:1.3
Niacin mg	0.053	0.001	53:1
Pantohentic Acid mg	0.753	0.08	9.4:1
Folacin ug	26	5	5.2:1
Vitamin E IU	0.51	0	

Essential Amino Acids	Abs value	Normalize	Abs value	Normalize	RDA	Norm RDA
Tryptophan	0.041	1	0.051	1	3	1
Threonine	0.151	3.7	0.149	2.9	8	2.7
Isoleucine	0.16	3.9	0.204	4.0	12	4
Leucine	0.237	5.8	0.291	5.7	16	5.3
Lysine	0.189	4.6	0.206	4.0	12	4
Methionine	0.071	1.7	0.13	2.5	10	3.3
Phenylalanine	0.121	3.0	0.21	4.1	16	5.3
Valine	0.17	4.1	0.251	4.9	14	4.7
Histidine	0.067	1.6	0.076	1.5	?	?

Source: Dunne, Lavon, Kirshmann, John. *Nutrition Almanac*. New York: McGraw-Hill Publishing Co. 1990.

A Summary of Some Protein Functions

Enzymes	Functions

Digestive Enzymes
- Amylase
- Sucrase
- Proteases
- Lipases

- Starch → Maltose
- Sucrose → glucose + fructose
- Proteins → peptides
- Triglycerides → monoglycerides

Transferases
- Transaminase
- Carnitine transferase

- Exchange of amino groups
- Exchange of a fatty acid

Antioxidants
- Superoxide dismutase
- Glutathione peroxidase
- Catalase

- Destruction of superoxide
- Destruction of peroxide
- Destruction of peroxide

Redox
- Dehydrogenases
- Oxidases

- Addition or removal of 2 H
- Addition of oxygen

Polymerases
- DNA polymerase I

- Repair of DNA

Structural Proteins
- Keratin
- Collagen

- Hair, nails, hoofs, horns
- Connective tissue: tendons, etc.

Hormones and Neuropeptides
- Insulin
- Glucagons
- Oxytocin
- Vasopressin

- Activates glucose absorption
- Activates glucose release
- Causes uterine contraction
- Causes vascular contraction

Storage Proteins
- Casein
- Ferretin

- Milk protein—binds minerals
- Serum protein—binds iron

Transport Proteins
- Hemoglobin
- Myoglobin
- Serum albumin
- VLDL, LDL, HDL
- Carnitine carriers

- Transports oxygen in blood
- Transports oxygen in muscle
- Transports fatty acids in blood
- Transports lipids in blood
- Mitochondrial fatty acid transport

Contractile Proteins
- Actin
- Myosin

- Thin contractile muscle protein
- Thick contractile muscle protein

Receptor Proteins
- Insulin Receptors
- LDL Receptors

Immunological (Defense) Proteins
- Gamma Globulins

- Antibodies

Protein Fates

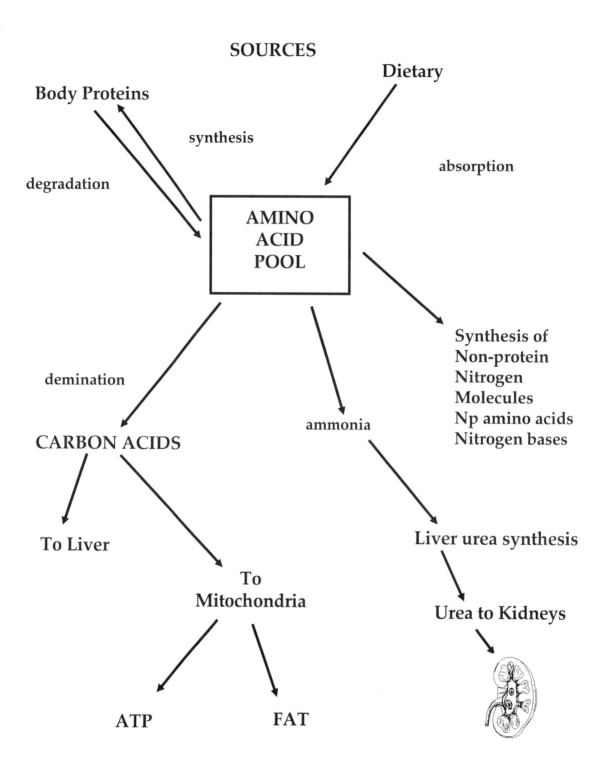

Nutritional Analysis
Phase II
Concept Map

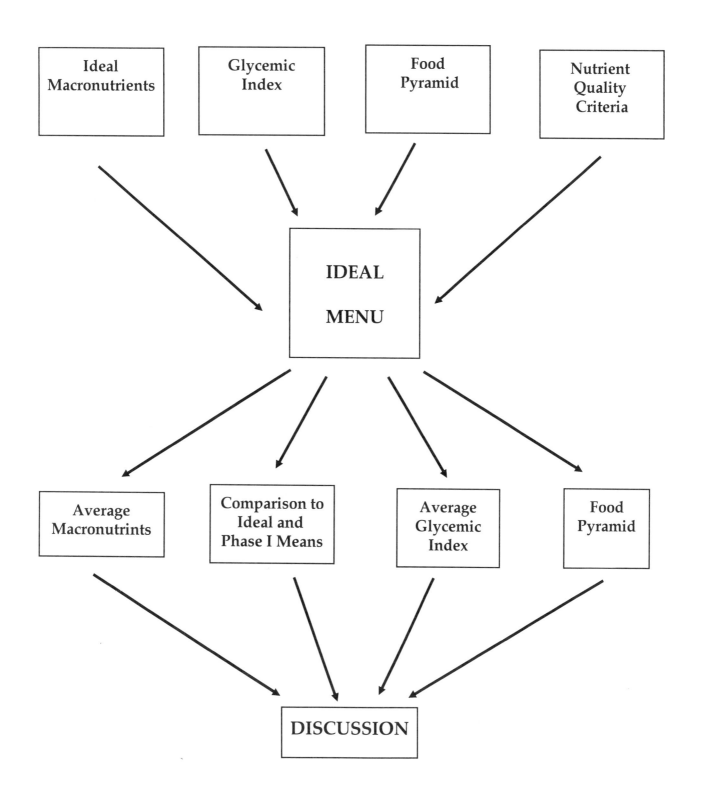

Nutritional Analysis
Phase II

The Synthesis of an Ideal Menu Analysis Is Health Status

Introduction

The nutritional analysis which you have completed delineates areas of over-nutrition and under-nutrition as you compared your average values to a set of either ideal values (macronutrients) or RDA values (in the case of micronutrients). These data are academically stimulating, but not very practical in terms of daily living. You need to be able to apply the results of your analysis by changing your own nutritional status to a more ideal state. The synthesis of an ideal daily menu is one way to accomplish this goal. This can lead you to an optimal state of nutritional health, which will maximize the probability that you will reach your genetic potential for physical and mental development.

The Synthesis of an Ideal Menu

You will prepare a list of foods with their amounts to be consumed for a **hypothetical** five-day period which optimizes quality and variety and whose macronutrient averages agree with your macronutrient values. Start choosing the foods that you like, then evaluate their nutritional quality using the criteria below. Use the food pyramid as a guide for the selection of quantities (servings) of various foods you choose. Eliminate the low quality foods by substituting high quality replacements. Now enter the food lists into the computer, so that you can obtain an analysis. After you enter each day, compare the total macronutrients on the screen to your set of properly calculated ideals from Phase I. Make adjustments in the amounts of the chosen foods or the food choices themselves until your totals approximately match those of the ideals. Continue to the next day, etc. until you have five days. You will need to print out a daily food list, the daily analysis, and the five-day averages.

Criteria Used to Judge the Quality of a Nutrient

- Energy—High quality kcals are obtained from foods which have a high nutrient density.
- Protein—High quality protein are those sources which: 1) have high nutrient density with respect to protein, 2) have high efficiency, 3) have a high protein/fat ratio, 4) have a high nutrient density with respect to micronutrients, 5) can be easily digested.

- Carbo—High quality carbos are those which have: 1) a high carbohydrate/kcal ration, 2) a high nutrient density, 3) a high fiber/kcal ration, and 4) a low glycemic index value.
- Fat—High quality fats are those which have: 1) a good balance between saturated, monounsaturated and polyunsaturated fatty acids (e.g., 1:1:1 or 1:1), 2) a high nutrient density, and 3) have both EFAs present in appropriate amounts.
- Fiber—High quality fiber foods are those which have a high nutrient density and a high fiber/kcal ratio. Ideally your ratio of insoluble to soluble fiber is 3:1.

When you begin this synthesis, it is important that you select foods that you like so that you are likely to eat them. Your menu design must:

1. Maximize adherence to food triangle recommendations and maximize variety of food choices.

2. Maximize adherence to ideal macronutrient values.

3. Maximize nutrient quality (use criteria above).

Once you have a rough selection of foods, you will enter the food lists into the computer, one day at a time, so that deletions and additions can be made until your macronutrient values match the ideals within, plus or minus 20 percent. Then you will prepare a table which compares the macronutrient averages with your calculated ideal values used in Phase I. You will prepare a table which lists your daily carbohydrate sources and their glycemic index values, which you can obtain from the glycemic index handout. You will calculate an average glycemic index for each day and for the five days.

Discussion

You will write a 600-word (minimum) discussion which must include all of the following **in order**:

1. A rationale for food choices. Why did you choose the foods?

2. How you chose foods from the food triangle to maximize variety and optimize adherence to the food triangle recommendations.

3. Did you maximize nutrient quality with respect to each of the five-macronutrient categories? **Which judgment criteria** did you use in determining nutrient quality? (**list them**). Did you choose any low quality food (use criteria list); and if so, why? How did your average glycemic index value turn out? If it is not moderate to low, what changes could you make to lower it?

4. Did you create a menu which will optimize your nutritional status? Explain. Could you honestly follow such a menu?

5. How does this ideal menu, that you have created, correct the problems which were evident in your Phase I analysis?

6. A discussion of the assessment lines for exercise, stress, blood chemistry and genetics.

7. A conclusion which creatively states the outcome of this diet.

Check List

Point distribution (You must have all of the following in the **following order!**)

1. Title page written according to course guidelines.

2. A set of daily food lists with the five-day average computer analysis. (10 pts)

3. A table comparing your average values to your calculated ideal values from Phase I. (5 pts)

4. A table comparing your carbohydrate glycemic index values and daily average. (5 pts)

5. Display of all four-line continuums. (5pts)

6. A discussion (minimum 600 words typed) which includes all of the criteria listed in the instructions for discussion. (25 pts)

7. References cited. List all references used.

Note: all rules and regulations for paper published in the HUN 1201 Course Guidelines must be followed.

Analysis of Health Status

Your health status and risk for the development of degenerative disease depends on multiple factors, including: 1) nutritional status, 2) exercise status, 3) stress management status, 4) blood chemistry risk factors, and 5) genetics. Nutritional status is being addressed thoroughly. Your exercise status can be assessed by placing yourself on the "Risk Factor Analysis" on the following page:

Risk Factor Analysis

Aerobic Exercise Continuum

Never Exercise	Seldom Exercise	Often Exercise	Regularly Exercise 3–4 times/wk

Stress Management Assessment Continuum

Usually cannot cope	Often cannot cope	Usually can cope	Almost always can cope

Blood Chemistry Risk Factors Continuum

Have high blood Cholesterol Glucose LDL Uric acid Blood pressure	Have not had my blood tested	Have normal blood Cholesterol Glucose LDL Uric acid Blood pressure

Genetic Risk Factors Continuum

One or more relatives(s) has had cancer heart disease diabetes stroke other degenerative disease	A distant relative has had cancer heart disease diabetes stroke other degenerative disease	I do not know	No relatives have had any degenerative disease

Quiz

1. Name a plant food combination that provides complete protein.

   _____

2. The food source that has the highest protein to fat ratio is

   _____.

3. The food source that has the highest nutrient density with respect to protein is

   _____.

4. Substances like lipase and superoxide dismutase are proteins called

   _____.

5. Proteins that act as chemical messengers are called

   _____.

6. An example of a chemical messenger protein

   _____.

7. Contractile proteins in muscles are, for example, called

   _____.

8. An example of a transport protein

   _____.

9. Gamma globulins are proteins called

   _____.

10. Connective tissue is composed of a protein called

   _____.

Quiz

1. Different amino acids are distinguished by their

    _____.

2. The amino acid cysteine is unique because it forms a _____ bond between parts of a protein.

    _____

3. The amino acid sequence of a protein is held together by

    _____.

4. A few amino acids in a chain comprise an entity called a/an

    _____.

5. The protein source which has the highest protein to fat ratio is

    _____.

6. The three dimensional shape of a protein dictates its

    _____.

7. When a protein loses its three dimensional shape it is said to be

    _____.

8. The bond denoted by, $-\overset{\overset{\displaystyle O}{\|}}{\underset{\underset{\displaystyle H}{|}}{C}}-N-$, is called a/an

    _____.

9. Excess protein (amino acids) in the diet leads to the production of a toxic waste product called

    _____.

10. Human protein contains _____ different amino acids.

    _____

Quiz

1. List the functional groups found in every amino acid molecule.

   _____

   _____

2. Name one amino acid which is not used to make protein.

   _____ .

3. The number of amino acids in a tripeptide

   _____ .

4. The number of dietary essential amino acids

   _____ .

5. Protein's three-dimensional shape is vital to the protein's cellular

   _____ .

6. The protein source which has the highest protein to fat ratio

   _____ .

7. The food source which has the highest nutrient density with respect to protein

   _____ .

8. When two or more amino acids combine chemically, they form a bond called

   _____ .

9. When you compare domestic meat sources to their wild meat equals, the ratio of protein/fat is lowest in the

   _____ .

10. What combination of plant protein sources can provide all of the essential amino acids?

   _____ .

CHAPTER 6
METABOLISM

Class Notes

Review of Cellular Organelles in which Metabolism Occurs

- Nucleus—synthesis of DNA and RNA
- Rough endoplasmic reticulum with ribosomes—protein synthesis
- Smooth endoplasmic reticulum—synthesis of complex carbohydrates and lipids
- Cytoplasm—Glycolysis
- Mitochondria—Catabolism—ATP synthesis

Central Metabolic Pathway

- The merging of the carbohydrate, lipid, and protein pathways.

Glycolysis

- Glucose → Pyruvate
- Mitochondrial catabolism
- Pyruvate → Acetyl coenzyme A
- Acetyl groups to the Kreb's cycle (citric acid, TCA cycle)
- Hydrogens to the electron transport chain (system)—ATP made
- Glucose is necessary in the diet so that oxaloacetate can be made; otherwise, the citric acid cycle does not function.

Electron Transport System (Chain)

- NADH carries hydrogens to the ETS.
- Hydrogens are received by a flavo protein containing FMN and iron.
- Hydrogens are passed on to coenzyme Q_{10}.
- Hydrogens are released as H^+.
- Electrons from the hydrogen are passed to cytochrome b, then to cytochrome c, then to cytochrome a, a_3, then to oxygen to make water.
- The energy from the electrons is used to make ATP.

Fatty Acid Cycle (Beta Oxidation Cycle)

- Fatty acids are activated by combining with coenzyme A.
- The activated fatty acids enter the fatty acid cycle and are converted to keto acids. Each turn of the cycle produces another acetyl coA.
- The fatty acid cycle can run in reverse in which acetyl coAs enter the cycle, then are combined to form saturated fatty acids. Cholesterol synthesis is a spin-off of this cycle.

Protein

- Amino acids are synthesized by transamination and amination.
- Amino acids are destroyed by deamination.
- Amino acids are synthesized into protein by ribosomes, following the template of mRNA.

Feasting Leads to:

- Maximum glycogen stores
- Increased fat stores
- Excess amino acids
- Adequate amino acids for protein synthesis

Fasting Leads to:

- Glycogen depletion
- Fatty acid mobilization

Extended Fasting Leads to:

- Gluconeogenesis
- Production of ketoacids via the fatty acid cycle
- Muscle wasting to liberate amino acids for other protein synthesis and gluconeogenesis

Metabolism

- Anabolism
  - Putting together SNMs to make the compounds of life.
  - Amino acids → protein and non-protein nitrogen molecules (nitrogen bases, amine neurotransmitters)
  - Glucose → Glycogen
  - Fatty acid → Triglycerides, Phospholipids

- Catabolism
    - Product=Hydrogens. Waste=CO_2
    - Glycolysis—glucose → Pyruvate
    - Citric Acid (TCA, Kreb's) Cycle—acetyl in, hydrogens and carbon dioxide (CO_2) out.
    - Cycle that receives acetyl groups from Acetyl Co A. Product=Hydrogens. Waste=CO_2.
    - ETS—ATP—Hydrogens in and water and ATP out. Produce=ATP. Waste=Water.
    - Fatty Acid Cycle—Acetyl CoA in, Fatty acids out (saturated).
        - Coenzyme A carries acetyl groups into fatty acid cycle, carries fatty acids into fatty acid cycle.
        - Makes saturated fatty acids, and cholesterol.
    - Urea Cycle—ammonia + carbon dioxide in → urea out.
        - Nitrogen waste product from amino acids.
    - Cholesterol—no catabolism, converted to bile in liver, then secreted.
    - Gluconeogenesis—amino acid → glucose.

The Pyruvate—Oxaloacetate Connection

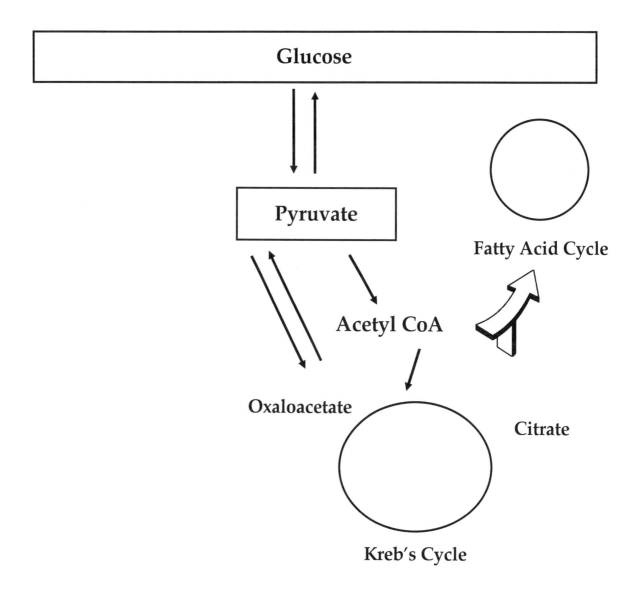

EFA Anabolism (Prostaglandin Formatin)

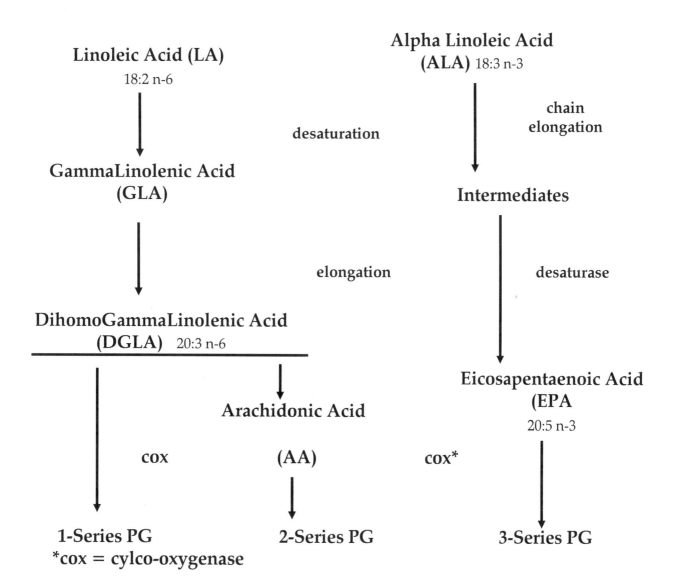

Linoleic Acid (LA)
18:2 n-6

Alpha Linoleic Acid
(ALA) 18:3 n-3

desaturation

chain elongation

GammaLinolenic Acid
(GLA)

Intermediates

elongation

desaturase

DihomoGammaLinolenic Acid
(DGLA)   20:3 n-6

Arachidonic Acid

Eicosapentaenoic Acid
(EPA
20:5 n-3

cox

(AA)

cox*

1-Series PG
*cox = cylco-oxygenase

2-Series PG

3-Series PG

Quiz

1. Energy metabolism occurs in the cellular organelle called

   _____.

2. The three major metabolic pathways join to form a

   _____.

3. A major product of carbohydrate anabolism

   _____.

4. Carbohydrate glycolysis converts glucose to

   _____.

5. When acetyl groups form in metabolism, they are picked up and carried by the coenzyme

   _____.

6. In the TCA (citric acid cycle) the carbons from acetyl groups are converted to

   _____.

7. The main hydrogen carrier molecule is

   _____.

8. Hydrogens are transported to the energy pathway called

   _____.

9. When fatty acids are catabolized, they are fed into the cycle called

   _____.

10. The anabolism of amino acids results in the formation of

    _____.

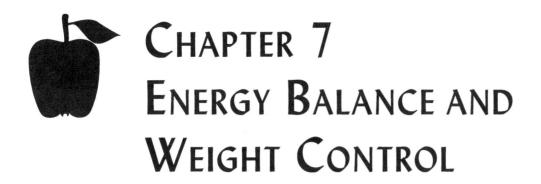

CHAPTER 7
ENERGY BALANCE AND
WEIGHT CONTROL

Class Notes

ATP Cycle

Bomb Calorimeter

Energy Expenditure

- BMR—basal metabolic rate
- TEF—thermic effect of food
- Adaptive thermogenesis
- Physical activity

Energy Balance

- $Kcals_{in} = Kcals_{out}$

Weight Gain

- $Kcals_{in} > Kcals_{out}$

Weight Loss

- $Kcals_{in} < Kcals_{out}$

Percent Body Fat

- Men (15–18%)
- Women (20–25%)

Factors That Influence Hunger, Food Choices, and Eating Habits

Body Mass Index

BMI Related to Mortality

Adipose Storage in the Body

Hormonal Control of Body Fat

- Leptin
- Resistin

Proliferation of Adipose Tissue

Energy Balance and Weight Control

- BMR—all body functions working while you are at rest.
- Leptin—hormone produced when fat is put into fat tissue. Suppresses appetite.
- Thermic effect of food—involved in energy production, heat production, the calories that we produce in the processing of foods.
- Active/Reactive Thermogenesis—Function of aerobic exercise. The higher the aerobic exercise, the higher Thermogenesis.
- Physical Exertion—burns kcals. Very important in having energy balance.
- Weight loss diets—main problem with weight loss is that you gain back the weight you lost. High fat, low carbohydrate diets can lead to gluconeogenesis.
- Exercise.

Methods of Determining Body Fat

Risk Factor

- Men > 25% (15-18%)

- Women > 30 (20-25%)

- Underwater weighing

- Skinfold Thickness

- Bioelectric Impedance Analysis

- High Technology Methods

  - CAT Scans

  - MRI Scans

OB Gene

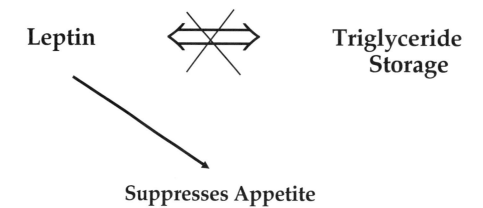

Feasting |——————————————————— Inhibitors

Leptin   **Triglyceride
Storage**

Suppresses Appetite

The Protein Resistin Is Associated with Type II Diabetes and Obesity

- Resistin protein is discovered in mice.

- Resistin gene is discovered in mice.

- When resistin protein is increased in mice, type II insulin resistance develops in two days.

- Resistin protein and gene discovered in humans (Univ. of Pennsylvania).

- Resistin receptor site has not been found. Strong drug implication.

- Monoclonal antibody has been synthesized.

Lazar, Dr. Mitchell, "The Hormone Resistin links obesity to diabetes." *Nature.* 409:307-12. January 18, 2001.

CHAPTER 8
WATER SOLUBLE VITAMINS

Class Notes

Vitamin C—Ascorbic Acid

- Functions
  - Chelating agent for Fe and other minerals
  - Increases Fe absorption
  - Antioxidant—fat-soluble form protects PUFAs
  - Free radical scavenger
  - Synthesis of collagen, incorporation of proline
  - Deficiency causes bleeding gums, bruising, immune compromise, scurvy
- Sources—fruits and vegetables
- Synergisms—B vitamins and other antioxidants
- Antagonists—oxidation, heat, light, stress, alcohol, smoke, low nutrient density

Thiamin—B-1

- Functions
  - Coenzyme in oxidative decarboxylation (alphaketo acids)
  - Vital to nervous system's synthesis of neurotransmitters
  - Thiamin is poorly stored and can be depleted in the body rapidly (days)
- Sources—pork, wheat germ
- Synergisms—B_1 through B_6 + Biotin exhibit an energy synergism
- Antagonists—high refined carbs and/or high fat diet, excessive alcohol consumption causes malabsorption syndrome for most water vitamins, but particularly thiamin, which results in irreversible brain and heart damage. See Vitamin C.

Riboflavin—B-2

- Function—forms coenzymes FMN and FAD—transports hydrogen
- Sources—mushrooms, spinach, broccoli, animal foods
- Synergisms—B_1 through B_6 + Biotin exhibit an energy synergism
- Antagonists—light, see vitamin C

Niacin—B-3

- Function—forms coenzymes NAD+ and NADP+—Transports hydrogen
- Sources—mushrooms, wheat bran, animal foods
- Synergisms—B_1 through B_6 + Biotin exhibit an energy synergism
- Antagonists—see vitamin C

Pantothenic Acid—B-5

- Function—forms coenzyme A—transports acid fragments
- Sources—mushrooms, broccoli, and most non-processed foods
- Synergisms—B_1 through B_6 + Biotin exhibit an energy synergism
- Antagonists—see vitaminC

Pyridoxine, Pyridoxal, Pyridoxal Phospate—B-6

- Function—reactions involving amino acids and PUFA biosynthesis into pro-staglandins
  - Protein metabolism
  - Red blood cell synthesis
  - Neuerotransmitter synthesis
  - Nucleic acid synthesis
- Sources—spinach, broccoli, banana, fish
- Synergisms—B_1 through B_6 + Biotin exhibit an energy synergism
- Angagonists—see vitamin C

Biotin

- Function—transfer of carbon dioxide to molecules, chain elonation lipid biosynthesis
- Sources—raw mushrooms, broccoli, most animal and plant foods
- Synergisms—B_1 through B_6 + Biotin exhibit an energy synergism
- Antagonists—heat, raw egg white, see vitamin C

Folacin, Folate, Folic Acid—B-9

- Functions—synthesis of RNA, DNA and red blood cells
- Protection of chromosomal fragile sites
- Sources—green leafy vegetables
- Synergisms—cyanocobalamin
- Antagonists—food procession, see vitamin C

Cyanocobalamin—B-12

- Functions—Synthesis of RNA, DNA, and hemoglobin
- Antipernicious anemia vitamin. Most stored water-soluble vitamin.
- Sources—animal foods
- Synergisms—folacin
- Antagonists—see vitamin C

Vitamins

General Characteristics

- Composed of organic molecules that we cannot synthesize in the body.

- Usually combined with other molecules to produce coenzymes.

- Work in synergistic groups (e.g., Folacin and Cyanocobalamin).

- Mostly removed during food processing.

- Many factors either destroy or increase the demand on vitamins:

  - Heat

  - Light (especially riboflavin)

  - Oxidation

  - Stress

  - Alcohol (especially thiamin)

  - Smoking

  - Drugs

  - Low nutrient density diet

  - Excessive physical activity

Vitamins (Vita = Life   Amin = Amines)

Water Soluble

- Vitamin C—Ascorbic acid (ate)

- Vitamin B-1—Thiamin

- Vitamin B-2—Riboflavin

- Vitamin B-3—Niacin

- Vitamin B-5—Pantothenic acid (ate)

- Vitamin B-6—Pyridoxal

- Vitamin B-9—Folacin, Folic acid (ate)

- Vitamin B-12—Cyanocobalamin, Biotin

Fat Soluble

- Vitamin A—Retinal, Retinol

- Vitamin D—Cholecalciferol

- Vitamin E—D-Alphatocopherol

- Vitamin K—Phylloquinone

Functions of Thiamin

Energy Metabolism as TPP

- Decrease in energy production

- Development of metabolic acidosis

Brain and Neurological Malfunction

- Due to lack of carbohydrate metabolism

- Development of mental confusion

- Poor memory

- Spastic gait

Metabolism of Some Amino Acids

Thiamin

Coenzyme in Decarboxylation

Carboxylic Acid $\rightarrow$ acid fragment and carbon dioxide

$$R - C \overset{O}{\underset{OH}{\big|}} \quad \xrightarrow[\text{TPP}]{\text{Decarboxylase}} \quad R\text{-} \ + \ CO_2$$

TPP = Thiamine Pyrophosphate

Example:

$$\text{Pyruvate} \xrightarrow{\text{Decarboxylase + TPP}} \begin{array}{c}\text{Acetyl CoA +}\\ \text{carbon dioxide}\end{array}$$

CoenzA   NAD+   NADH + H$^+$

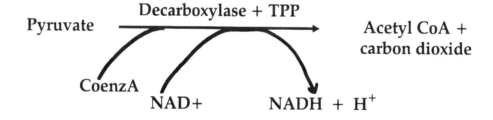

Sources of Thiamin by Nutrient Density

Food	Serving size to yield 1.5 mg (RDA)	Kcals to yeild 1.5 mg
Ham	5.5 oz	252
Wheat germ	1 cup	268
Canadian bacon	8 pieces	340
Sunflower seeds	1/2 cup	369
Green peas	3 1/4 cup	466
Pork chop	5.5 oz	468
Watermelon	4 slices	585
Oatmeal	5 cups	745

Functions of Riboflavin

Energy Metabolism as FMN and FAD

- Decrease in energy production

- Development of growth cessation

Riboflavin

Flavin + Ribitol = Riboflavin

Riboflavin + Phosphate = FMN (Flavin Mononucleotide)

FMN + AMP + FAD (Flavin-Adenine Mononucleotide)

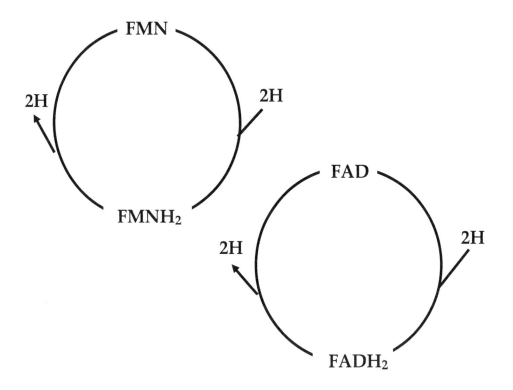

Sources of Riboflavin by Nutrient Density

Food	Serving size to yield 1.7 mg (RDA)	Kcals to yield 1.7 mg
Mushrooms, cooked	3 3/4 cups	158
Spinach, cooked	4 cups	164
Broccoli, cooked	5 1/3 cups	243
Non-fat milk	4 cups	360
1% milk	4 cups	410
Buttermilk	4.5 cups	440
Oysters	4 cups	640
Whole milk	4 cups	648

Functions of Niacin

Energy Metabolism as NAD+ and NADP+

- Decrease in energy production

- Development of neurological and muscular malfunction

Synthesis from Tryptophan

- Tryptophan → Niacin

Sources of Niacin by Nutrient Density

Food	Serving size to yield 10 mg	Kcals needed for 19 mg
Mushrooms, cooked	2.5 cups	109
Wheat bran	1.75 cups	140
Tuna, canned	4 oz.	183
Chicken breast, roasted	0.75	214
Salmon, pink canned raw	8 oz	317
Halibut, broiled	7 oz	326

Functions of Pantothenic Acid

Energy Metabolism as Coenzyme A

- HS—ethylamine-pantothenic acid-ADP-P

The Acid Fragment Carrier

- Examples:

    - Fatty Acid Cycle

    - Citric Acid Cycle

    - Glycolysis—Pyruvate → AcetylCoa

Sources of Pantothenic Acid by Nutrient Density

Food	Serving size to yield 2.3 mg	Kcals needed for 2.3 mg
Mushrooms, cooked	1.5 cups	27
Broccoli, cooked	2.33 cups	114
Lobster, cooked	4.7 oz	211
Eggs, cooked	3.7	213
Milk, 1%	3 cups	270
Chicken, roasted	1.5 cups	331

Functions of Pyridoxal B-6

Energy Metabolism as PLP

- Decrease in immune function

- Development of metabolic acidosis

Protein and Amino Acid Metabolism

- Functions in:

    - Protein synthesis

    - Non-protein amino acid synthesis

    - Amination

    - Deamination

    - Transamination

    - Amino acid decarboxylation

- Examples:

  - Tyrosine → dopa → dopamine →   norephinephrine → epinephrine

  - Histidine → histamine

  - Tryptophan → serotonin

- Deficiency Symptoms

  - Development of weakness

  - Failure to grow and repair tissue

  - Impaired motor function

Unsaturated Fatty Acid Metabolism

- PUFAs → eicosanoids → prostaglandins

- PMS

Sources of Pyridoxal by Nutrient Density

Food	Serving size to yield 2.0 mg	Kcals needed for 2.0 mg
Spinach, cooked	4.5 cups	186
Broccoli, cooked	6.33 cups	292
Banana, raw	2 cups	318
Salmon, broiled/baked	8.5 oz	440
Watermelon	2.75 slices	440
Chicken breast, roasted	2 whole	600
Cantaloupe	3 each	605
Tuna	13.5	614

Functions of Biotin

Transfers Carbon Dioxide to Molecules (Creating Carboxyl Groups)

Chain Elongation of Polyunsaturated Fatty Acids

Sources of Biotin by Nutrient Density

Cauliflower

Egg Yolk

Functions of Folacin

Functions

- Synthesis of purines (adenine, gluanine)

- Synthesis of pyrimidines (cytosine, thiamine, uracil)

- Synthesis of red blood cells (RBC)

- Prevents chromosomal mutations

Deficiency Manifestations

- Spinal bifida—open spine

- Cleft palate

- Cervical cancer

- Prostate cancer

- Other cancers

- Other teratogenic effects

Sources of Folate by Nutrient Density

Food	Serving size to yield 200 ug	Kcals needed for 200 ug
Spinach, raw	1.75 cups	22
Romaine lettuce	2.5 cups	23
Spinach, cooked	0.75 cup	31
Turnip greens, cooked	1 cup	34
Asparagus, cooked	1 cup	48
Collard greens, cooked	1 cup	57
Broccoli, cooked	1.75 cups	86
Beets, cooked	2 cups	104
Orange juice	1.75 cups	205
Wheat germ, raw	0.75 cup	219
Cantaloupe	1.25 each	235

Functions of Cyanocobalamin

Synthesis of Purines (Adenine, Guanine)

Synthesis of Pyrimidines (Cytosine, Thymine, Uracil)

Synthesis of Hemoglobin (Pernicious Anemia)

Folate Activation

Fatty Acid Cycle

Sources of Vitamin B-12 by Nutrient Density

Food	Serving size to yield 2 ug	Kcals needed for 2 ug
Oyster, raw	1 oz	7
Clams, fried	0.5 oz	10
Crab, Alaskan King	2 oz	17
Bluefish, broiled	3.5 oz	51
Beef, filet	2.5 oz	164

Functions of Vitamin C

Chelating Agent for Fe and Other Metals

Enhances Iron (Fe) Absorption

Antioxidant (#1)

Free Radical Scavenger

Synthesis of Collagen

Activates the Immune System—T-Lymphocytes

Deficiency

- Bleeding gums

- Easy bruising

- Immune compromise

- Scurvy

Sources of Ascorbic Acid by Nutrient Density

Food	Serving size to yield 100 mg	Kcals needed for 100 mg
Green peppers, raw	1.66 each	20
Cauliflower	1.66 cups	45
Bok choy cabbage	2.5 cups	45
Croccoli, cooked	1.66 cups	45
Strawberries, fresh	1.25 cups	55
Brussel sprouts, cooked	1.66 cups	60
Papaya, fresh	0.5 cup	65
Romaine lettuce	7.5 cups	70
Grapefruit, red	2.5 each	80
Orange, fresh, medium	1.66 each	85
Asparagus, cooked	2.5 sups	90
Orange juice, fresh	1 cup	90

Quiz

1. The active form of vitamin A is

    _____ .

2. The plant substance that is known as pre-vitamin A is

    _____ .

3. The rate of cell division is partially controlled by

    _____ .

4. The fat-soluble vitamins which are found mostly in plant foods are vitamins

    _____ .

5. The #1 antioxidant protector of polyunsaturated fatty acids in cell membranes is
    known as vitamin

    _____ .

6. Primary electrolytes include the metal ions

    _____ .

7. The best source (by nutrient density) of potassium is

    _____ .

8. The most nutrient dense source of sodium and chloride is

    _____ .

9. The primary electrolytes help maintain acid-base balance and _____
    balance.

10. The antisterility vitamin for males is

    _____ .

Test 3 Outline

- Protein

    - Structure

    - N balance $N_{in} = N_{out}$

    - Enzymes

    - Antibodies

    - Hormones

    - Receptor sites

    - Transporters

- Metabolism

    - Anabolism

        - Amino acids → protein and non-protein nitrogen molecules (nitrogen bases, amine neurotransmitters)

        - Glucose → glycogen

        - Fatty acid → triglycerides, phospholipids

    - Catabolism

    - Glycolysis—glucose → pyruvate

- Citric acid (TCA Kreb's) cycle—acetul in, hydrogens and carbon dioxide out

- ETS—ATP—hydrogens in, and water and ATP out

- Fatty acid cycle—acetyl CoA in, fatty acids out (saturated)

- Urea cycle ammonia + carbon dioxide in → urea out

- Cholesterol—no catabolism, converted to bile in the liver, then secreted

- Gluconeogenesis amino acid → glucose

- Energy Balance and Weight Control

  - BMR

  - Leptin

  - Thermic effect of food

  - Reactive thermogenesis

  - Physical exertion

  - Weight loss diets

  - Exercise

- Water Soluble Vitamins

  - Thiamine—ham, pork

    - TPP—decarboxylation (nerve function)

- Riboflavin—mushrooms

    - FMN, FAD hydrogen carriers

- Niacin—mushrooms

    - #1 hydrogen carrier NAD+

- Pantothenic acid—mushrooms

    - Coenzyme A

- Pyridoxal—spinach

    - Amino acid and protein metabolism

    - Metabolism of PUFAs

- Folacin, Folic acid, Folate—spinach

    - Chromosome protection

    - Synthesis of nitrogen bases

    - Synthesis of RBC

- Cyanocobalamin—oysters

    - Synthesis of nitrogen bases

    - Synthesis of hemoglobin

- Biotin—cauliflower

    - Chain elongation

    - Ascorbic acid—peppers

- Antioxidant

    - Collagen synthesis

    - Immune function

Exam Review Sheet—The Bottom Line

Vitamins (Vita=Life   Amin=Amines)

General Characteristics

- Composed of organic molecules that we cannot synthesize in the body.
- Usually combine with other molecules to produce coenzymes.
- Work in synergistic groups (working together to enhance the outcome) (folacin and cyancobalamin).
- Mostly removed during food processing.
- Many factors either destroy or increase the demand on vitamins.
    - Heat
    - Light (especially riboflavin)
    - Oxidation
    - Stress
    - Alcohol (especially thiamin)
    - Smoking
    - Drugs (birth control pills)
    - Low nutrient density diet
    - Physical activity

Folacin and cyanocobalamin are one example of a synergistic relationship; they work together to enhance the making of protein, RNA and DNA. The other water-soluble vitamins work together in what we call energy metabolism. There are groups of vitamins that work together as antioxidants. Vitamins are mostly removed during food processing. The enrichment program adds the following lost micronutrients back, B-1, B-2, B-3, B-9, and iron. We lose all the other B vitamins. We can wind up with deficiencies even though they're "enriched."

Many factors either destroy or increase the demand on vitamins. Destroy means that they are chemically wiped out. Increase the demand means that we either destroy them inside the body or we use them up so that our demand is increased. Heat is a destroyer, especially in the case of water-soluble vitamins. The higher the heat the more complete the destruction. If you fry or broil food, you do more destruction than if you boil something, because boiling is limited to the boiling point of water (100° C); if you fry something the temperature on the surface of the pan is 400° F. Studies show that microwave cooking is the best way to cook food in terms of preserving the vitamin content—except B-12. Somehow, microwaves destroy B-12, which we only get from animal products. If you're careful you can cook your animal products in the microwave.

Light destroys vitamins, especially riboflavin. Vitamins should be kept in an opaque light resistant container. Oxidation can occur just from the oxygen in the air. Vitamins should be kept in a sealed container in the refrigerator. If you do not keep vitamins in the refrigerator they will decompose. Oxidation can occur from the air, getting into the vitamins or by free radicals in the body. Stress increases the demand for vitamins. Alcohol destroys vitamins, especially thiamin. Thiamin is very sensitive to alcohol and is literally destroyed by it. The destruction of thiamin leads to neurological destruction and delirium tremors. Smoking is the single worst thing you can do to destroy a vitamin. Smoking is a concentration of free radicals; they attack everything. They destroy many things, including protein, molecules, receptor sites, and vitamins. Any drug can destroy a vitamin, so if you are taking a prescription drug you need to take some kind of supplement. Low nutrient density diets increase the demand for supplementation. Physical activity puts a tremendous demand on vitamins and increases the need for supplementation as well. We literally destroy or flush micronutrients out of the body during physical activity. The more active you are the more vitamins you destroy.

Vitamins work as coenzymes that work with the enzyme proteins to cause the chemical reactions to go very fast at a rate that's compatible with LIFE.

Three things you need to know about vitamins:

- What is the functional form of this vitamin?
- What does it do? (What kind of chemical reactions does it get involved in?)
- What is the number-one source?

Thiamin B-1

- Number one food source is ham

Functions of Thiamin

- Energy Metabolism as TPP
  - Decrease in energy production
  - Development of metabolic acidosis
- Brain and neurological malfunction
  - Due to lack of carbohydrate metabolism
  - Development of mental confusion
  - Poor memory
  - Spastic gait
- Coenzyme in decaraboxylation
- Carboxylic acid $\rightarrow$ acid fragment and CO_2
- Example: pyruvate $\rightarrow$ decarboxylation + TPP $\rightarrow$ amino acids and CO2

TPP = thiamin pyrophosphate is the active form of thiamin. Thiamin functions in energy metabolism as TPP. TPP is the coenzyme in decarboxilation. Breaks up the acid group.

Produces acid fragment carbon dioxide. Chops CO_2 off with TPP. Example: pyruvate through acetyl coA plus carbon dioxide so we're using the TPP to chop off the CO_2. This leaves the acid fragment which is picked up by coenzyme A, which is the acid fragment carrier. So whenever we produce an acid fragment we need to have coenzyme A around. So coenzyme A picks up the acetyl group. NAD^+ picks up hydrogen. So we produce $NADH + H^+$. TPP is used in the carbohydrate pathway, kreb cycle and some amino acid pathways to remove carbon dioxide.

Reminder: Enrichment adds thiamin, riboflavin, niacin, folate and iron.

Riboflavin

Functions of Riboflavin B-2

- Energy Metabolism as FMN and FAD
  - Decrease in energy production
  - Development of growth cessation

Flavin + Ribitol = Riboflavin

Riboflavin + Phosphate = FMN (Flavin Mononucleotide)

FMN + AMP = FAD (Flavin Adenine)

- Two nucleotides put together (nitrogen base, sugar, phosphate)
- Di nucleotide = Flavin mononucleotide + Adenine mononucleotide (AMP)
- Flavin is a nitrogen base
  - Glycogen depletion
  - Fatty acid mobilization

Extended Fasting Leads To:

- Gluconeogenesis
- Production of ketoacids via the fatty acid cycle
- Muscle wasting to liberate amino acids for other protein synthesis and gluconeogenesis

Niacin functions in the active form, niacinamide. Used as a dinucleotide coenzyme, NAD^+, with dehydrogenase enzymes. NAD is the number one hydrogen carrier in metabolism. NAD = niacinamide adenine dinucleotide. The number one source by nutrient density is mushrooms.

Pantothenic acid—pantothenate is a component in the coenzyme A complex. Coenzyme A (CoA) is the acid fragment carrier. It functions in the fatty acid cycle, the citric acid cycle, and the conversion of pyruvate to acetylCoA. The number one source by nutrient density is mushrooms.

Pyridoxal is the active form of vitamin B-6. It is used as the shuttle for amino acids in metabolism. In this capacity it functions as a coenzyme with the enzyme systems that make amino acids, convert one amino acid into another, as well as those that synthesize protein. It is also used as a coenzyme in the anabolism of polyunsaturated fatty acids to prostaglandins. The number one source if spinach. Supplement in excess of 50 mg/day can be neurotoxic.

Folacin (folic acid, folate) is a vitamin coenzyme that works synergistally with cyanocobalamin. Folate is activated by the presence of cyancobalamin. They function together in the synthesis of the nitrogen bases of DNA and RNA, i.e., the purines (adenine and guanine) and pyrimidines (cystone, thymine, and uracil). Folate helps in the production of red blood cells. It also functions in the metabolism of homocysteine, which has been implicated in the process of atherogenesis, leading to coronary artery disease (CAD). The most recent accolade for folate is its role in protecting the fragile sites of chromosomes from mutation. This is one of the leading causes of birth defects as well as adult mutation leading to increased aging and cancer. The number one source by nutrient density is spinach. All of the leading sources are plants, mostly green plants.

Cyanocobalamin is a large vitamin molecule, whose absorption is facilitated by a gastric secreted protein called the *intrinsic factor*. Cyanocobalamin (B-12) works synergistically with folate in the production of purines and pyrimidines. Together they function in growth, repair and reproduction of all tissues in the body. Cyanocobalamin functions in the production of hemoglobin. The number one source by nutrient density is oysters. It is found in all animal foods, but it is not found in plants.

Biotin is a vitamin molecule that functions as a coenzyme in the transfer of carbons. Chain elongation of fatty acids requires biotin. Adding or deleting carbons from carbohydrate or amino acid molecules requires biotin. It is found in all natural food, thus deficiencies are unknown. The number one source by nutrient density is cauliflower.

Ascorbic acid, vitamin C, is a delicate, water soluble molecule, that is obtained from fruits and vegetables. The number one source by nutrient density is green peppers. Ascorbate is a potent antioxidant. Studies show that the concentration of the potent oxidation product of lipid peroxidation, malondialdehyde, is low in patients that have taken supplements of vitamin C. Ascorbate is a potent activator of T-lymphocytes, white blood cells that destroy viruses. It functions synergistically with zinc in this capacity. Ascorbate also functions as a coenzyme with iron in the formation of the connective tissue, collagen. It helps maintain strong capillaries; deficiencies lead to bleeding due to capillary fragility. Ascorbate is a good chelating agent; thus it enhances the absorption of minerals, especially iron.

CHAPTER 9
FAT SOLUBLE VITAMINS

Class Notes

Retinal—Vitamin A (Beta Carotene)

- Sources—Liver (beta carotene: yellow/orange fruits and vegetables, dark green vegetables)
- Functions
  - Beta carotene —vitamin A required Zn
  - Night vision
  - Mucopolysaccharide production
  - Immune function—T-cell activity
  - Bone growth and development
  - Antioxidant in cell membranes—anticarcinogen
- Toxicity—beta carotene safe, retinal and retinoic acid (retin-A) are toxic in excessive doses

Vitamin D—Cholecalciferol, D_3; Ergocalciferol, D_2

- Sources—dairy products (typically enhanced) and oily fish, sunshine vitamin (cholesterol—vitamin D cascade)
- Functions
  - Stimulates cells to produce calcium-binding proteins
  - Regulation of calcium/phosphorus metabolism in cooperation with the parathyroid glands
  - Bone growth and calcification
- Toxicity—very toxic—causes excessive calcium deposits

D-Alphatocopherol—Vitamin E

- Sources—vegetable oils (nuts, seeds, whole grains, asparagus)
- Functions
    - Antioxidant, free radical scavenger, GSHP, SOD
    - Cell membrane protection—PUFA protection—RBCs are especially sensitive
    - Prostaglandin synthesis

Vitamin K

- Sources—leafy green vegetables
- Functions—regulatory factor in the clotting mechanism

Fat Soluble Vitamins

Vitamin A—Retinol, *Retinal*, Retinoic Acid (Retin A)

- Sources: Yellow/orange fruits and vegetables

- Functions

  - Vision—night blindness

  - Protein synthesis and cell differentiation

  - Cell division

  - Beta carotene is an antioxidant

  - Demineralization of bone in injury

Vitamin D—Cholecalciferol, Plant Version Ergocalciferol

- Sources: Animal sources—eggs, milk, fish

- Functions

  - Calcium absorption and transportation

  - Calcium deposition in bone

Vitamin E—D-Alphatocopherol (200–400 IU/Day) as Mixed Tocopherols

- Sources: vegetables oils (grains, nuts), seeds, sweet potato

- Functions

    - Antioxidant—#1 protector of PUFAs

    - Antisterility in males

    - Coenzyme in the synthesis of prostaglandins

Vitamin K—Phylloquinone

- Sources: leafy green vegetables

- Function

    - Clotting factor

The Fat Soluble Vitamins

Retinol (retinal) is a fat soluble vitamin (vitamin A) that comes from yellow/orange fruits and vegetables as the pre-vitamin, beta carotene. Retinal functions in the nerve sensitive retina of the eye, aiding in night vision. It is a regulator of cell division, thus contributing to the control of growth, repair and reproduction of tissue. Serves as a coenzyme in liver protein synthesis, bone demineralization and growth, skin growth and repair, and adrenal gland function. Both retinal and beta carotene act as antioxidants in the blood and in cell membranes. Supplements of 100,000 iu/day can lead to serve overdose and death.

Cholecalciferol (vitamin D) is known as the sunshine vitamin because it is made from cholesterol when ultraviolet light in incident on the skin. In addition to the in vivo de novo synthesis, vitamin D is obtained from animal sources in their fat, e.g., egg yolk, seafood and fortified milk. Vitamin D functions in the absorption, transportation, and metabolism of calcium and phosphorus. It acts as a coenzyme in liver, kidney and bone function. The strength of bones and teeth depend on the adequate presence of vitamin D. Huge overdoses can lead to calcium deposits in soft tissues including skin, kidneys, and blood vessels.

D-Alphatocopherol (vitamin E) is a fat soluble vitamin that is the principal protector of polyunsaturated fatty acids in cell membranes. It is synergistic with the other antioxidant factors. It functions as an antioxidant in the blood stream as well as cell membranes. Vitamin E is known as the antisterility vitamin in males. It acts as a coenzyme in the prostrate gland and is a component of prostatic fluid, which supports the viability of sperm. It is a coenzyme in the production of prostaglandins, which are internal cell regulators. Vitamin E has antithrombolytic activity, thus can prevent coronary thrombosis and stroke. Research has also confirmed that vitamin E has anticarcinogenic activity. Nuts, seeds, grains, legumes and vegetables are all sources.

Phylloquinone (vitamin K) is a fat soluble vitamin that is used as a coenzyme in the synthesis of blood clotting factors. It is also used in the liver in various metabolic reactions. Leafy green vegetables are the chief source. Colon flora synthesize it.

Fat Soluble Vitamins Equivalents Chart

Vitamin	RDA	Conversions
A (Retinal)	1500 RE	5000 IU
D (Cholecalciferol	400 IU	10 ug
E (D-Alphatocopherol)	8 mg/10mg	30 IU = 9 mg
K (phylloquinone	80 ug	

Quiz

1. The active form of vitamin A is

   _____ .

2. The plant substance that is known as pre-vitamin A is

   _____ .

3. The rate of cell division is partially controlled by

   _____ .

4. The fat soluble vitamins which are found mostly in plant foods are vitamins

   _____ .

5. The #1 antioxidant protector of polyunsaturated fatty acids in cell membranes is known as vitamin

   _____ .

Chapter 10
Minerals

Class Notes

Sodium

- Electrolyte and fluid balance, nerve function via depolarization

Potassium

- Nerve impulse transmission, electrolyte balance
- Imbalance causes cardiac arrhythmias—Na/K pump in cell membranes

Chloride

- Negative electrolyte along with HCO_3, HCl formation in stomach

Calcium

- Absorption problems, vitamin D, estrogen, parathyroid hormone, and calcitriol facilitate absorption
- Functions
  - Blood clotting
  - Nerve/muscle transmission (tetany)
  - Calmodulin cell metabolism regulating system (activates cells)
  - Bone calcification age 25–30 (strength—osteoporosis)

Phosphorus

- Calcium phosphate (hydroxyapatite) bone deposition
- Phosphates activate many essential biochemicals, ATP

Magnesium

- 60 percent in bone
- Green vegetables and whole grains best sources

- Functions
  - Coenzyme in many (over 300) reactions (ATP cycle)
  - Proper nerve and cardiac function
- Deficiency can cause sudden heart attack (athletes, distance runners)

Iron

- Absorption problems—aided by vitamin C
- Functions
  - Hemoglobin
  - Cytochromes in ETS
  - Coenzyme with vitamin C in collagen synthesis, coenzyme
- Anemia, iron storage disease (ferritin, hemosiderin)

Zinc

- Oysters, seafood, chicken, beef, legumes
- Functions
  - Over 200 enzyme systems require Zn: Cu/Zn SOD
  - Carbonic anhydrase
  - Alcohol dehydrogenase
  - Immune system (T-cell activity)
  - Vitamin A mobil from liver
  - Sexual function (prostrate)
  - Prostaglandin synthesis
  - Nucleic acid and protein metabolism
- TOXIC over 35 mg/day. Zn/Cu

Copper

- Oysters, pecans, lobster
- Functions
  - Coenzyme
  - ETS
  - Iron absorption and metabolism
  - Needed to form collagen network strength
  - SOD

Selenium

- GSHP, selenoamino acids, cell membrane PUFA protection, synergistic with vitamin E and beta carotene, most effective free radical scavenger
- Deficiency leads to muscle wasting, muscle pain and heart disease
- Sources: Shrimp, seafood best sources

Iodine

- Thyroxine

Chromium

- Activates insulin, deficiency leads to diabetes and heart disease
- Sources: oily vegetables, egg yolks, whole grains

Manganese

- Mitochondrial SOD, cereals, vegetables

Molybdenum (MO)

- Grains and legumes
- Functions
  - Sulfite oxidase
  - Xanthine oxidase (Xanthine → Uric Acid)
  - Aldehyde oxidase

General Characteristics of Minerals

- Minerals are naturally occurring elements on earth.

- They cannot be destroyed by heat, light, or chemical action.

- They can be dialyzed out of food or the human body by excessive water.

Absorption Is Enhanced By:

- Heme (meat)

- Acid conditions

- Ascorbic acid

Absorption Is Inhibited By:

- Fiber

- Phytates (grain)

- Oxalates (spinach, rhubarb)

- Tannins (teas and herbs)

- Mineral interactions

- Basic pH

Excellent Mineral Supplements

Mineral	Amout	Source
Potassium	K 25 mg	Any source—both are very water soluble
Calcium	Ca 1200 mg	Calcium aspartate
Magnesium	Mg 600 mg	Magnesium aspartate Ascorbates are good
Iron	Fe 18 mg	Iron (ferrous) aspartate Ferrous ascorbate Ferrous gluconate
Zinc	Zn 15 mg	Zinc picolinate
Copper	Cu No supplement recommended	
Manganese	Mn 10 mg	Manganese aspartate
Molybdenum	Mo 10 mcg	Molybdenum aspartate
Chromium	Cr 100 mcg	Chromium picolinate
Selenium	Se 100 mcg	Selenoamino acid

Functions of Sodium

Primary Electrolyte in Blood Stream (135–145 meq/l)

Fluid Balance and Acid Base Balance

- Nerve Transmission

- Muscle Contraction

Imbalance Will Cause Heart Attack (<130, >145)

Functions of Potassium
K^+

Primary Electrolyte in Blood Stream (3.5–4.5 meq/l)

Fluid Balance and Acid Base Balance

- Nerve Transmission

- Muscle Contraction

In Glycolysis—Pyruvate Kinase (Mg^{++})

Protein Biosynthesis—Also Requires Mg^{++}

Imbalance Will Cause Heart Attack (<3.5, >4.5)

Food	Kcals	Potassium	ND K
Beet greens	8	208	26.0
Endive	8	158	19.8
Chinese cabbage	9	176	19.6
Celery	18	340	18.9
Spinach	14	259	18.5
Pickles dill	11	200	18.2
Parsley	26	436	16.8
Radish	7	104	14.9
Cauliflower	24	356	14.8
Mushrooms	18	260	14.4
Molasses	**43**	**585**	**13.6**
Rhubarb	26	351	13.5
Asparagus	30	404	13.5
Vege Juice	**41**	**535**	**13.0**
Kohlrabi	38	490	12.9
Tomato	**46**	**552**	**12.0**
Broccoli	24	286	11.9
Pumpkin	**49**	**564**	**11.5**
Cucumber	14	156	11.1
Squash, summer	25	263	10.5
Kale	33	299	9.1
Turnips	39	348	8.9
Cantaloupe	**94**	**825**	**8.8**
Peppers chili	30	255	8.5
Peppers, sweet	24	196	8.2
Casaba	43	344	8.0
Okra	38	302	7.9
Collard greens	35	275	7.9
Sauerkraut	42	329	7.8
Honeydew	46	350	7.6
Carrots	48	356	7.4
Squash, winter	129	945	7.3
Yams	**210**	**1508**	**7.2**
Artichoke	65	434	6.7
Papaya	**117**	**780**	**6.7**
Apricot	51	313	6.1
Parsnips	**102**	**587**	**5.8**
Guava	45	256	5.7
Strawberries	45	247	5.5
Kiwi fruit	46	252	5.5
Lima beans	**208**	**969**	**4.7**
Mulberries	61	271	4.4
Gooseberries	67	297	4.4
Grapefruit	38	167	4.4
Banana	105	451	4.3
Plantain	**181**	**739**	**4.1**
Orange	62	237	3.8
Dates	**228**	**541**	**2.4**

Functions of Calcium

Coenzyme for Blood Clotting Reactions

Excites Muscular Activity, Especially the Heart

Creates Bone Density—Calcium Phosphate

Coenzyme in a Number of Reactions

On Switch for Cellular Metabolism

Sources of Calcium

Food	Kcals	Calcium	ND Ca
Cabbage, China	9	74	8.2
Watercress	7	53	7.6
Collard greens	35	218	6.2
Beet greens	8	46	5.8
Parsley	26	122	4.7
Rhubarb	26	105	4.0
Spinach	14	51	3.6
Yogurt, lowfat	127	452	3.6
Endive	8	26	3.3
Molasses	43	137	3.2
Milk, skim	96	302	3.1
Cheese, Parm	111	336	3.0
Chard, Swiss	6	18	3.0
Cheese, Roman	110	302	2.7
Kale	33	90	2.7
Cheese, Swiss	107	272	2.5
Cheese, Moz, LF	72	183	2.5
Lettuce, romaine	8	20	2.5
Cheese, Gruyer	117	287	2.5
Celery	18	44	2.4

Functions of Magnesium

Coenzyme for All ATP Reactions

Calms Muscular Activity, Especially the Heart

Coenzyme in the Synthesis Reactions for DNA and RNA

Coenzyme in Phosphate Transfer Reactions

Off Switch for Cellular Metabolism
("Nature's Tranquilizer" Adele Davis)

Synthesis of Protein

Coenzyme in the Synthesis of Prostaglandins

Coenzyme in over 300 Reactions

Sources of Magnesium

Food	Kcals	Magnesium	ND Mg
Chard, Swiss	6	30	5.0
Beet greens	8	28	3.5
Spinach	14	44	3.1
Bran, wheat	**121**	**279**	**2.3**
Okra	38	56	1.5
Molasses	43	52	1.2
Pickles, dill	11	12	1.1
Endive	8	8	1.0
Watercress	7	7	1.0
Peanut lowfat	**223**	**216**	**1.0**
Parsley	26	25	1.0
Pumpkin	**774**	**738**	**1.0**
Artichoke	65	60	0.9
Broccoli	24	22	0.9
Collard greens	35	31	0.9
Cucumber	14	12	0.9
Squash, summer	25	21	0.8
Wheat germ	**431**	**362**	**0.8**
Beans, green	34	27	0.8
Oysters	58	46	0.8
Halibut	93	71	0.8
Scallops	75	48	0.6

Creatine

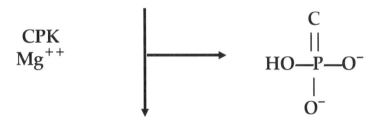

Creatine Phosphate

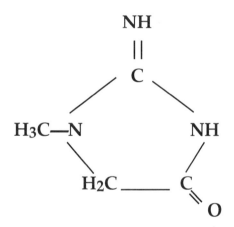

(burned during short bursts of activity)

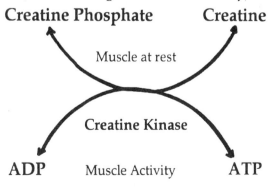

A Coupled Reaction

Functions of Iron

Central Ion in HEME

Functions in Hemoglobin, Myoglobin, Cytochromes (ETS), and Iron-Sulfer Protein (ETS)

Coenzyme in Many Reactions

Coenzyme in the Synthesis of Collagen

Iron Storage Disease (Hemachromatosis) Can Be Fatal

Sources of Iron

Food	Kcals	Iron	ND Fe
Total cereal	**116**	**21**	**0.18**
Beet greens	8	1.2	0.15
Parsley	26	3.7	0.14
Spinach	14	1.7	0.12
Oat flakes	**177**	**13.7**	**0.08**
Cream of Wheat	**134**	**10.3**	**0.08**
Molasses	43	3.2	0.07
Puffed corn cereal	110	8.1	0.07
Onion, green	26	1.88	0.07
Tofu	72	5.2	0.07
Wheat bran	121	8.5	0.07
Peppers	24	1.2	0.05
Potato	**386**	**18.9**	**0.05**
Tomato juice	46	2.2	0.05
Artichoke	114	5.1	0.04
Mulberries	61	2.6	0.04
Oat Cherrios	111	4.5	0.04
Venison	**572**	**22.7**	**0.04**
Snails	117	4.3	0.04
Kale	33	1.14	0.03
Leeks	76	2.6	0.03
Collard greens	35	1.16	0.03
Broccoli	24	0.78	0.03
Brussels sprouts	38	1.2	0.03
Abalone	89	2.7	0.03
Beef, dried	47	1.3	0.03

Functions of Zinc

Activates the Immune system (T-Lymphocytes)

Coenzyme in Ethanol Dehydrogenase

Coenzyme in Carbonic Anhydrase

Coenzyme in Delta Desaturase (Prostaglandin Synthesis Pathways)

Sources of Zinc

Food	Kcals	Zinc	ND Zn
Bran, wheat	121	279	2.3
Oysters	58	76	1.3
Crab	71	5	0.1
Collard greens	35	1.8	0.1
Sauerkraut	42	1.9	0.05
Lobster	77	2.6	0.03
Asparagus	30	0.94	0.03
Clams	133	2.5	0.02
Yogurt, low fat	127	2.2	0.02
Chick, dk w/o skin	136	2.2	0.02
Black beans	116	1.41	0.01
Cheese, gouda	101	1.11	0.01
Cheese, edam	101	1.1	0.01
Cheese, Swiss	101	1.11	0.01

Functions of Manganese

Mitochondrial SOD

Superoxide → water

.O₂-

Sources of Manganese

Food	Kcals	Manganese	ND Mn
Syrup, corn	57	2.8	0.05
Sauerkraut	42	1.9	0.05
Pineapple	77	2.6	0.03
Okra	38	0.99	0.03
Parsley	26	0.56	0.02
Collard greens	35	0.69	0.02
Kale	33	0.52	0.02
Boysenberries	66	0.72	0.01
Asparagus	30	0.29	0.01
Green beans	34	0.24	0.01
Artichoke	65	0.43	0.01
Oysters	58	0.38	0.01
Trout	126	0.72	0.01
Black beans	116	0.64	0.01
Persimmon	118	0.6	0.01

Functions of Molybdenum
"Big Mo"

Coenzyme in Aldehyde Oxidase

- Aldehyde → acid

Xanthine Oxidase

- Xanthine → uric acid

Sulfite Oxidase

- Sulfite → sulfate

- Sulfite can cause anaphylactic shock reaction

Sources of MO

Beans

Peas

Functions of Iodine

Integral Component of Thyroxine

#1 Source = Seafood

Functions of Chromium

Activates Pro-Insulin

May Be Involved with Glut 4 and/or Insulin Receptors

Involved with Lipid Metabolism—
Perhaps in the Shaping of Carnitine Transferase Protein

#1 Source = Mushrooms/Asparagus

Functions of Selenium

Coenzyme for Glutathione Peroxidase

Forms Selenoamine Acids in which Se Acts as an Antioxidant

Involved with Some Transport Proteins

Bonds with Toxic Heavy Metals—Decreasing Store Amounts in the Liver and Other Tissues

#1 Source = Seafood

Benefits of Aerobic Exercise

- Increases the thermic effect of food

- Lowers resting heart rate

- Increases the number of cellular receptor sites

- Increases muscle mass

- Decreases fat stores

- Stimulates production of endorphins

- Enhances the body's immune status

- Decreases the onset of degenerative diseases (atherosclerosis, cancer, diabetes)

- Decreases LDL and increases HDL

- Optimizes cardiopulmonary function (heart and lungs work more efficiently)

- Increases capillaries throughout the body

- Increases metabolic rate, thus increasing energy

- Increases the size and number of mitochondria

- Increase the body's efficiency of removing wastes

Nutritional Recommendations for PMS

Two Weeks Before Your Period Take the Following Daily Supplements

- 35 mg of zinc (picolinate or ascorbate)

- 600 mg of magnesium (aspartate or ascorbate)

- 50 mg vitamin B-6

- Per package directions—Evening Primrose Oil (in capsules)

- 2 capsules Max EPA (marine lipids)

- 400 IU vitamin E (micellized)

Do Not Eat

- Sweets

- Domestic Meats

- Fatty Foods

Carcinogens

Nitrates and Nitrates

Hydrogenated Oils

Smoking

Ionizing Radiation

Saturated Fat

Valium

Estrogens

Aflatoxins (Molds) (Peanut Butter)

Food Additives

Air and Water Pollution

Radon

Atherogenesis and Heart Disease

Conditions for Atherogenesis

- Serum cholesterol over 150 mg/dl

- High fat diet 30–40% of kcals as fat

- Serum triglycerides 100 mg/dl

- LDL over 120 mg/dl

- Statistically, HDL mg/dl

- Elevated serum VLDLs

Factors Related to Atherogenesis

- Sedentary lifestyle

- Antioxidant deficiency

- Poor genetic history

- Substance abuse

- Slow transit time

- Low plant fiber ingestion

1991 Known Etiology of Atherosclerosis

Elevated Levels of LDL Invade Arterial
Endothelium

LDL Oxidation Occurs Attracting
Monocytes

Monocytes Differentiate into
Macrophages—Begin Engorging
Oxidized LDL amd VLDL

toxic LDL

Enlarged Macrophages Are Damaged by
Oxidized Lipoproteins Causing Them to
Become Foam Cells (contain oxidized
triglycerides and cholesterol esters)

Foam Cells Die Leaving Cell Fragments
and Lipid Oxidation Products

Oxidation Products Damage
Endothelium Attracting Platelets to the
Lesion Which Causes Thrombogenesis

Outline for Final Exam

- Essential nutrients

- Nutrient classes

- Basic food groups and the food pyramid

- Relative abundance of nutrients in body

- Food advertising

- Nutritional research design—double blind, placebo controlled

- Enrichment program

- Food processing—nutrient content, vitamins and minerals

- Calorie distribution—protein, carbohydrate, fat

- Appetite control—appestat (hypothalamus)

- Water soluble vitamins

- Fat soluble vitamins

- Major minerals

- Trace minerals

- Balanced nutrients in nutrient classes

    - Sat. fat/monounsat fat/pufa

    - N_{in} vs. N_{out}

    - $Kcals_{in}$ vs. $Kcals_{out}$

    - Vitamins/minerals

- Food labels

- Nutrient density

- United States recommendations for dietary status

- "Fattening"—excess kcal intake, overproduction of insulin

- Factors that influence insulin production

- How to maintain a relatively constant insulin level

- Complex carbohydrate vs. simple carbohydrates

- Reactive hypoglycemia

- Diabetes I and II

- Glycemic index

- Glucose and the CNS

- Glucagon secretion/function

- Gluconeogenesis—occurs during carbohydrate deprivation

- Fiber—soluble vs. insoluble—sources, functions

- Fatty acids—saturated, monounsaturated, polyunsaturated (pufas)—structure, sources

- Triglycerides

- EFAs 18:2 n-6, 18:3—sources

- Cholesterol—uses, problems, disposal (soluble fiber), sources

- Transport of lipids, chylomicrons, LDL, VLDL, HDL

- Phospholipids, cell membranes, micelles

- Lipid hydrogenation

- Lipid oxidation, cell membrane damage, origin of degenerative disease

- Amino acids—structure, essential amino acids, non-protein amino acids

- Peptides, peptide bonds, amino acid chains

- Protein structure dictated by DNA via mRNA

- Ribosomes—protein factories

- Protein function—enzymes, hormones, receptor sites, transport, acid-base balance, fluid balance

- Protein efficiency

- Complete protein

- Digestion of carbohydrate, protein, fat

- Digestive secretions—saliva, gastric juice, bile, pancreatic juice, intestinal juice

- Digestive enzymes—names and function

- Villi, brush border cells

- Absorption mechanisms—diffusion, facilitated diffusion, active transport

- Metabolism

  - Catabolism—small nutrient molecules to wastes + ATP

  - Anabolism—small nutrient molecules to the compounds of life

- Carbohydrate metabolism pathways

  - Glycolysis (glucose to pyruvate)—pyruvate to acetyl CoA—Citric acid cycle electron transport chain

- Lipid metabolism pathways

  - Fatty acids—fatty acylCoA—fatty acid cycle (beta oxidation cycle—acetyl CoA]

- Anabolism—Acetyl CoA to fatty acid cycle—saturated fat (cholesterol is also made

- Excess cholesterol to HDL to liver to bile

- EFAs to prostaglandins

- Leptin

- Protein metabolism

  - Protein to amino acids deamination to acids to the carbohydrate pathway

  - Nitrogen waste—amino groups to ammonium to the liver, then to urea to kidneys

- Pathogenesis of atherosclerosis—high cholesterol, high LDL, oxidation of LDL

- Cancer development—mutation, promotion, progression and infiltration—metatasis

- Antioxidants—Vitamins C, E and A, beta carotene—enzymes glutathione peroxidase (Se), superoxide dismutase (Zn and Mn), catalase (Zn)

- Pnytonutrients—cruciferous vegetables

- Vitamin C—ascorbic acid—stimulates the immune system, collagen synthesis

- Vitamin E—D-Alphatocopherol—whole grains—antioxidant, pufa protector

- Vitamin A (plants have beta carotene+ 2 retinal)—night vision, cell division

- Vitamin D—calcium shuttle, calcium deposits in bone. Excess causes calcium deposits in soft tissue

- Iodine—seafood main source—thyroxine production

- Selenium—glutathione peroxidase

- Calcium—mostly in bone, blood clotting, #1 source spinach

- Potassium—primary electrolyte, nerve function, membrane polarization, #1 source spinach

- Sodium—#1 source salt. Primary electrolyte

- Magnesium—coenzyme in ATP reactions and DNA/RNA synthesis. #1 source—spinach

- Iron (heme)—used in hemoglobin, myoglogin, cytochromes (ETS)

Quiz

1. Primary electrolytes include the metal ions

    _____ .

2. The best source (by nutrient density) of potassium is

    _____ .

3. Calcium activates cell metabolic activity. It enters the cells through membrane proteins called

    _____ .

4. The metal ion that is a coenzyme for all ATP reactions and synthesis of RNA and DNA is

    _____ .

5. Americans are typically deficient in magnesium because they do not consume adequate amounts of

    _____ .